Editor's preface

Electron microscopy is now a standard technique with wide applications in all branches of Science and Technology, and every year a large number of students and research workers start to use the electron microscope and require to be introduced to the instrument and to the techniques for the preparation of specimens. Many books are available describing the techniques of electron microscopy in general terms, but the authors of Practical Methods in Electron Microscopy consider that there is an urgent need for a comprehensive series of laboratory handbooks in which all the techniques of electron microscopy are described in sufficient detail to enable the isolated worker to carry them out successfully. The series of books will eventually cover the whole range of techniques for electron microscopy, including the instrument itself, methods of specimen preparation in biology and the materials sciences, and the analysis of electron micrographs. Only well-established techniques which have been used successfully outside their laboratory of origin will be included.

Great care has been taken in the selection of the authors since it is well known that it is not possible to describe a technique with sufficient practical detail for it to be followed accurately unless one is familiar with the technique oneself. This fact is only too obvious in certain 'one author' texts in which the information provided quickly ceases to be of any practical value once the author moves outside the field of his own experience.

Each book of the series will start from first principles, assuming no specialist knowledge, and will be complete in itself. Following the successful innovation, made by the same publishers in the parallel series Laboratory

Techniques in Biochemistry and Molecular Biology (edited by T. S. Work and E. Work), each book will be included, together with one or two others of the series, in a hardback edition suitable for libraries and will also be available in an inexpensive edition for individual use in the laboratory. Each book will be revised, independently of the others, at such times as the authors and editor consider necessary, thus keeping the series of books continuously up-to-date.

Strangeways Research Laboratory
Cambridge, England

AUDREY M. GLAUERT, SC.D.
General editor

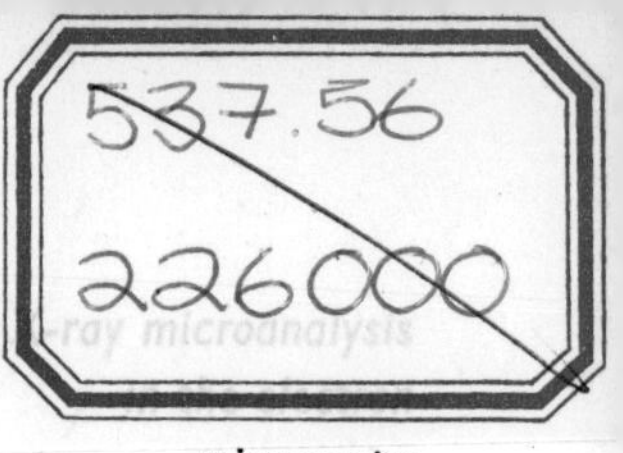

microscope

A

Practical Methods in ELECTRON MICROSCOPY

Edited by

AUDREY M. GLAUERT

Strangeways Research Laboratory

Cambridge

NORTH-HOLLAND PUBLISHING COMPANY
AMSTERDAM · NEW YORK · OXFORD

X-RAY MICROANALYSIS IN THE ELECTRON MICROSCOPE

JOHN A. CHANDLER
Tenovus Institute for Cancer Research,
Welsh National School of Medicine, Cardiff

NORTH-HOLLAND PUBLISHING COMPANY
AMSTERDAM · NEW YORK · OXFORD

1st edition 1977
2nd printing 1978

Published by:

ELSEVIER/NORTH-HOLLAND BIOMEDICAL PRESS
335 JAN VAN GALENSTRAAT, P.O. BOX 211
AMSTERDAM, THE NETHERLANDS

Sole distributors for the U.S.A. and Canada:

ELSEVIER/NORTH-HOLLAND INC.
52 VANDERBILT AVENUE
NEW YORK, N.Y. 10017

North-Holland ISBN series 0 7204 4250 8
This book 0 7204 0607 2

This book is the laboratory edition of Volume 5, Part II, of the series 'Practical Methods in Electron Microscopy'. Volume 5 of the series contains the following parts:

Part I Staining methods for sectioned material
by P. R. Lewis, D. P. Knight
Part II X-ray microanalysis in the electron microscope
by J. A. Chandler

Titles of earlier volumes published in this series:

Volume 1 Part I Specimen preparation in materials science
by P. J. Goodhew
Part II Electron diffraction and optical diffraction techniques
by B. E. P. Beeston, R. W. Horne, R. Markham
Volume 2 Principles and practice of electron microscope operation
by A. W. Agar, R. H. Alderson, D. Chescoe
Volume 3 Part I Fixation, dehydration and embedding of biological specimens
by Audrey M. Glauert
Part II Ultramicrotomy
by Norma Reid
Volume 4 Design of the electron microscope laboratory
by R. H. Alderson

Printed in The Netherlands

To my family

Acknowledgements

I am grateful to a large number of people who gave me their help in the compilation of this volume. Especial thanks go to Miss Janet Nichols and to Miss Barbara Towler for painstakingly typing the manuscript.

Dr. Theodore Hall of Cambridge reviewed the chapter on quantitation and gave helpful corrections and suggestions. The section on metallurgical preparation and standards was read by Dr. Gordon Lorimer of Manchester who also provided constructive criticism, and I am thankful to both.

The manuscript could not have been completed without the continuous cooperation of, and discussion with, my colleagues in the Tenovus Institute. I am particularly indebted to Professor Keith Griffiths and to the Tenovus Organisation for providing facilities from which much of the information was obtained.

A number of the figures have been reproduced with the permission of John Wiley & Sons, Inc., New York. Photographs of some instruments were kindly provided by manufacturers as indicated.

My thanks go especially to Dr. Audrey Glauert who edited the manuscript with such care and provided constructive guidance throughout its preparation.

I am doubly grateful to Mr. Ron Alderson for both reviewing the manuscript and for first introducing me to the trials and rewards of X-ray analysis several years ago during my sojourn with AEI.

JOHN A. CHANDLER

Contents

Chapter 1

Introduction

1.1 *The nature of X-ray microanalysis*

When an electron beam strikes a solid specimen a number of interactions occur, the most important of these being illustrated in Fig. 1.1. Electrons may be back-scattered from the front face of the specimen with little or no energy loss, or they may interact with surface atoms to produce secondary (low energy) electrons. Some electrons may be absorbed by the specimen with transfer of energy to heat and sometimes to light. Transmitted electrons may be unchanged in direction or scattered at different angles. Scattered electrons may be elastic (no energy loss) or inelastic (having lost some energy). If energy is transmitted to the specimen it may also result in the production of Auger electrons or X-rays. Each of these events can provide information about the specimen. However the interaction which is the subject of this book is the generation of X-rays by high energy electrons passing through a thin specimen ($< 10\ \mu m$) in the electron microscope. These X-rays carry information about the atoms within the specimen in the region being irradiated, and thus provide a means of correlating the ultrastructural information in the electron microscope image with chemical analyses of very small regions of the specimen. X-ray microanalysis makes use of the fact that atoms, when struck by electrons from an external source, yield X-rays which are characteristic of those atoms. Consequently, the X-rays can be used to identify and quantify the elements present. Suitable detectors, placed close to the specimen, collect the X-rays and the information thus obtained is displayed for immediate interpretation of the specimen composition.

The two techniques of electron-optical imaging and X-ray analysis were

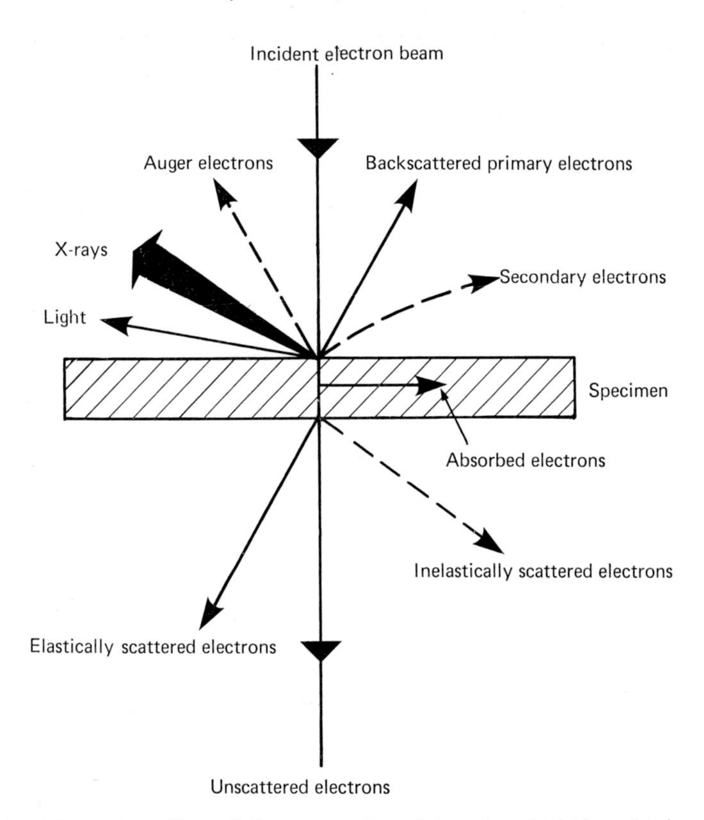

Fig. 1.1. The various effects of electron-specimen interaction. Light is emitted as visible fluorescence; elastically scattered electrons suffer no energy loss; inelastically scattered electrons lose some energy, and secondary electrons are of much lower energy than the primary electron beam.

first combined in the late 1940s by Castaing and Guinier (1949). Since then, and especially in the last 15 years, many hundreds of instruments with facilities for X-ray microanalysis have been produced and used in such diverse scientific areas as metallurgy, physics, electronics, mineralogy, environmental pollution and geology, and more recently in zoology, pathology, biochemistry and other biological fields.

This book describes the instruments and techniques involved in the analysis of relatively thin specimens, i.e. those specimens in which the full capabilities of electron microscopy are utilised. Application in both the transmission and scanning electron microscopes is discussed. The immensely large field of electron-probe microanalysis (mainly for bulk specimens) is not included except as an introduction to thin specimen analysis and the reader is referred to other literature for a detailed treatment of these techniques (see end of Chapter for list of references).

Throughout this book the emphasis is placed on practical aspects of analysis and only the minimum discussion of theoretical considerations is included. More detailed descriptions of the theoretical basis of microanalysis will be found in the references listed in Chapter 6. The information contained within this book is provided for both the beginner, with little or no experience in analytical techniques but with some knowledge of electron microscopy, and the more experienced operator who wishes to more fully understand some of the practical aspects of analysis in the electron microscope. Only well established techniques which have been used successfully outside their laboratory of origin are included.

1.2 *The philosophy of X-ray microanalysis*

Modern electron-optical instruments are able to provide resolving powers of less than 0.5 nm and the limitations to obtaining ultrastructural detail lie not so much in instrumental factors as in methods of specimen preparation, and in the techniques of image analysis available. The need also frequently arises to complement the available morphological information with a chemical analysis of certain regions of the specimen in order to determine the relationship between ultrastructural changes and variations in chemical composition of the specimen components. Without local chemical analysis or cytochemical techniques the electron microscopist can only interpret the image in terms of fine structure. For example the biologist, observing dynamic effects in a tissue, will want to know the relationship between the observed changes in structure and variations in elemental distribution throughout the material. Conversely the metallurgist, knowing the elemental composition of particular areas of his material will have a greater understanding of the morphological information presented by the microscope.

Clearly then there is a need to combine a method of obtaining high resolution images of thin specimens with simultaneous elemental analysis of a non-destructive nature of the same regions of the specimen. X-ray microanalysis fulfils these needs by providing an *in situ* means of identifying elements within microvolumes of thin specimens to a very high degree of sensitivity and with very precise localisation of the regions being analysed. This requires a combination of a high performance electron microscope with X-ray detectors that can be incorporated into the electron-optical system without compromising other facilities for specimen examination.

References

Castaing, R. and A. Guinier (1949), Sur l'exploration et l'analyse élémentaire d'un échantillon par une sonde électronique, Proc. 1st Int. Congr. Electron Microscopy, Delft, p. 60.

Further reading

The following references are suggested for further reading in general X-ray microanalysis, including analysis of bulk material.

Beaman, D. R. and J. R. Isasi (1972), Electron beam microanalysis, ASTM Special publication 506, (ASTM, Pennsylvania).

Birks, L. S. (1969), X-ray spectro-chemical analysis, 2nd edn. (John Wiley, New York).

Duncumb, P. and P. K. Shields (1963), The present state of quantitative X-ray microanalysis, part 1: Physical basis, Br. J. appl. Phys. *14*, 617.

Goldstein, J. I. and H. Yakowitz (1975), Practical scanning electron microscopy. Electron and ion microprobe analysis (Plenum Press, New York).

Hall, T. A. (1971), The microprobe assay of chemical elements, in: Physical techniques in biological research, 2nd edn., G. Oster, ed. (Academic Press, New York), p. 393.

Hall, T. A., P. Echlin and R. Kaufmann, eds (1974), Microprobe analysis as applied to cells and tissues (Academic Press, London and New York).

Hall, T. A., H. D. E. Rochert and R. L. de C. H. Saunders (1972), X-ray microscopy in clinical and experimental medicine (Ch. C. Thomas, Illinois, U.S.A.).

Heinrich, K. F. J. (1968), Quantitative electron probe microanalysis, NBS special publication 298, Washington D.C.

Liebhavsky, H. A., H. G. Pfeiffer, E. M. Winslow and P. D. Zemany (1960), X-ray absorption and emission in analytical chemistry (John Wiley, New York).

McKinley, T. D., K. F. J. Heinrich and D. B. Wittry (1966), The electron microprobe (John Wiley, New York).

Martin, P. M. and D. M. Poole (1971), Electron probe microanalysis: the relation between intensity ratio and concentration, Metals and Materials Metallurgical Reviews *5*, 19.

Philibert, J. (1970), Electron probe microanalysis, in: Techniques of metals research, Vol III, part 2, Modern analytical techniques for metals and alloys, R. Bunshah, ed. (Interscience, New York).

Poole, D. M. and P. M. Martin (1969), Electron probe microanalysis: instrumental and experimental aspects, Metallurgical Review, No. 133, 61.

Russ, J. C. (1972), Elemental X-ray analysis of materials (Edax International, Illinois, U.S.A.).

Various proceedings of The Electron Probe Society of America, (EPASA).

Various proceedings of The Illinois Institute of Technology, Research Institute, meetings on scanning electron microscopy. Ed. O. M. Johari, Chicago, Illinois.

Chapter 2

Production of X-rays

Every element in the periodic table has a very well-defined distribution of electrons within the atom. X-ray microanalysis is dependent on the excitation of these electrons to produce an emitted X-ray spectrum characteristic of the element concerned.

2.1 Model of the atom

Figure 2.1 represents a simple schematic representation of the atom. The nucleus, composed of protons and neutrons, is surrounded by electrons circulating in orbits. Each orbit corresponds quantum mechanically to a certain energy level of the electrons. The number of these orbits and energy levels depends on the size and state of the atom, and the orbits are grouped together into major units called shells. Elements are thus characterised by their nuclear charge and the energy distribution of their orbital electrons. Heavy elements, having large atoms, contain large numbers of electron orbits and shell units. In this simple model the shells nearest the nucleus are taken to contain electrons with the least potential energy. The shells are given the notation K, L, M etc. (Fig. 2.1) and in each shell there are a number of energy levels.

2.2 Ionisation of atoms

If one of the orbital electrons is in some way removed from its normal energy level, and perhaps ejected from the atom altogether, the atom is then in an excited state and is said to be ionised. In order to stabilise the atom an electron from a higher energy orbit falls immediately into this gap and its

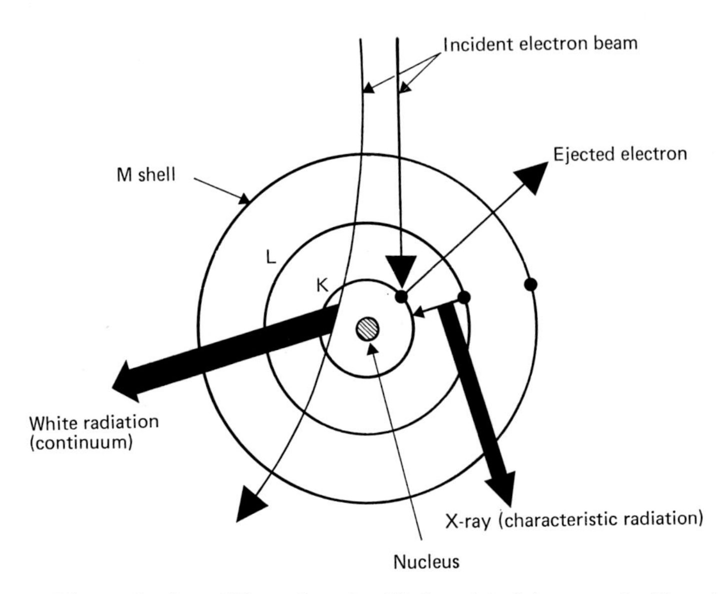

Fig. 2.1. The production of X-rays in a simplified model of the atom. Incident electrons must have sufficient energy to remove an orbital electron from its shell, leaving a space. Electrons falling into the space emit their excess energy as X-ray photons. White radiation is produced when the incident electron is decelerated in the field of the nucleus.

excess energy is emitted as an X-ray photon. This X-ray energy is the potential energy difference between the two shells. Thus, if the ejected electron comes from shell K (Fig. 2.1), and the gap is filled by an electron from shell L, the X-ray photon will have an energy $E_L - E_K$, where E_L and E_K are the respective electron-orbital energies. A vacancy will now be created in the L shell and another transition occurs almost simultaneously this time from, say, the M shell into the L shell, again giving rise to emission of an X-ray. In a large atom having a large number of electrons in orbit, a single ionisation event can thus give rise to a large number of orbital transitions and hence to a large number, or spectrum, of X-ray emissions.

2.3 Characterisation of elements

Some of the electron transitions that can occur in an atom when the inner K shell is ionised are illustrated in Fig. 2.2. A heavy element such as uranium ($Z = 92$) will give rise to a very great number of spectral emissions, whereas sodium ($Z = 11$) has only 11 orbital electrons and far fewer X-ray emissions. Not all transitions are permitted, however, by the laws of quantum mechanics.

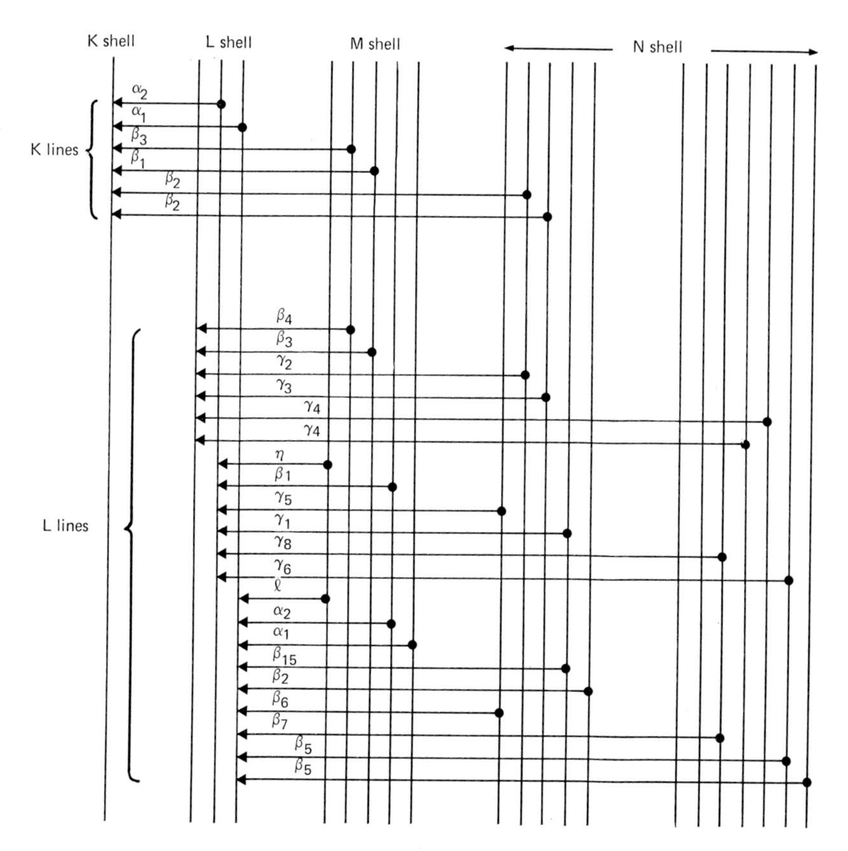

Fig. 2.2. Some X-ray emissions produced by electron transitions between orbits. Each shell possesses a number of electron orbits.

The emissions, or *characteristic X-ray lines*, vary also in intensity, and it is those with the greatest intensity that are most easily detected and are thus used for characterising the atom. In practice the most intense lines are those from the E_L–E_K transitions (called K lines), followed by those from E_M–E_L transitions (L lines), followed by those from E_N–E_M transitions (M lines) and so on. As Fig. 2.2 suggests, even within shells there are many possible electron jumps and these give rise to $K_{\alpha 1}$, $K_{\alpha 2}$, $K_{\beta 1}$, $K_{\beta 2}$, $K_{\gamma 1}$, $K_{\gamma 2}$ lines etc. in descending order of intensity.

There is a simple relationship between X-ray frequency ν (or energy) and atomic number Z given by

$$\nu = 0.248\ (Z{-}1)^2 \times 10^{16}$$

This equation was first proposed by Mosely in 1914 and since then all the elements of the periodic table have been characterised by this relationship and their X-ray emission spectra are well tabulated (Bearden 1964). Hence measurement of the energy of one of the X-ray emissions from an element allows that element to be identified. The frequency, v, of the X-ray radiation is related to its quantum energy E by the relationship $E = hv$ where h is Planck's constant. The relationship between wavelength, λ, and energy is given by $\lambda E = 12.4$.

2.4 Electron ionisation

X-ray microanalysis requires the generation of X-rays from a specimen by ionising the atoms within the specimen. In the electron microscope this ionisation is caused by the primary electron beam which must have sufficient energy to remove an electron from one of the inner shells of the atom concerned. This energy is known as the *critical excitation potential* or *absorption edge* and has a discrete value for each orbital electron energy level. The K electrons require greater excitation than the L electrons, and the energy levels differ between elements, generally increasing with atomic number. Figure 2.3 shows the energies of the X-rays emitted from elements of increasing atomic number, indicating the K, L and M lines and the most useful range of energies for X-ray detection. Typical X-ray emission energies

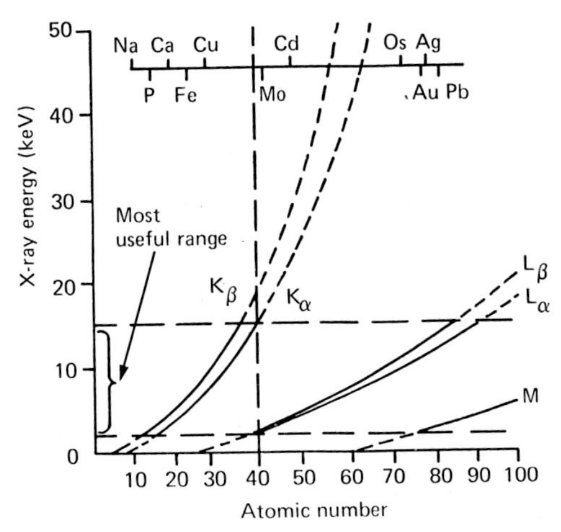

Fig. 2.3. X-ray energy of K, L and M lines versus atomic number. The most easily detectable range is from 2–15 keV and this may determine the choice of spectral line for analysis.

for a range of commonly detected elements of increasing atomic number are also shown in Table 3.2. Since the critical excitation potential is the energy required to remove the orbital electron from its shell, it is approximately equal to the total sum of the orbital energies outside of that shell. Thus the critical excitation potential (or absorption edge) for the K_α line of uranium (115.6 keV) is approximately equal to the sum of the $K_\alpha + L_\alpha + M_\alpha$ line energies ($98.4 + 13.6 + 3.2 = 115.2$) for that element.

2.4.1 X-ray yield

For an atom to be ionised the energy of the incident electron must be greater than the critical excitation potential of that atom. When X-ray photons are produced, however, they are not always totally emitted from the atom.

TABLE 2.1

Fluorescent yields for K lines of some elements

Atomic number	*Element*	*Fluorescent yield* (ω)
6	C	0.0009
7	N	0.0015
8	O	0.0022
10	Ne	0.0100
11	Na	0.020
12	Mg	0.030
13	Al	0.040
14	Si	0.055
15	P	0.070
16	S	0.090
17	Cl	0.105
18	Ar	0.125
19	K	0.140
20	Si	0.165
21	Ca	0.190
22	Ti	0.220
23	V	0.240
24	Cr	0.26
25	Mn	0.285
26	Fe	0.32
27	Co	0.345
28	Ni	0.375
29	Cu	0.41
30	Zn	0.435
40	Zr	0.73
50	Sn	0.86

There is a possibility of the X-ray photon being reabsorbed and producing *Auger electrons* (Burhop 1952). The probability of the photon being emitted is called the *X-ray fluorescent yield*, not to be confused with *X-ray fluorescence* (§ 2.4.2). The fluorescent yield increases with atomic number and is larger for K line emissions than L line emissions. The lower values of fluorescent yield for the low atomic number elements is one of the factors limiting their detection. A list of fluorescent yields is given by Birks (1969), and is shown for K_α radiation in Table 2.1.

It has been shown by Green and Cosslett (1961) that the efficiency of X-ray production from an element is a function of the fluorescent yield for that element, and of $(E_0 - E_c)^{1.63}$ where E_0 is the potential energy of the primary electron beam and E_c is the critical excitation potential associated with a particular X-ray line. The probability that an atom will be ionised by an electron of energy E_0 (called the *relative ionisation probability*) is given by $Q = (1/E_0\ E_c) \log_e (E_0/E_c)$. Q is also called the *ionisation cross-section.*

The maximum efficiency of X-ray production is obtained when the condition $E_0 \geqslant 2.7\ E_c$ is satisfied (see § 5.2.1). In practice the X-ray intensity from a sample is dependent on a number of factors including the number of atoms of the element concerned being irradiated, the atomic number, the probe voltage and the electron beam current.

2.4.2 X-ray fluorescence

As well as being ionised by incident electrons, atoms may be raised to an excited state by other X-rays provided the energy of those X-rays is high enough. Thus a primary electron beam may strike a specimen and cause atoms to be ionised and X-rays to be emitted. These X-rays, on passing out of the specimen, may interfere with other atoms and cause a secondary ionisation giving rise to further X-ray emissions. This effect is called *X-ray fluorescence* and can cause some errors in quantitation (see Chapter 6) since the additional fluorescent X-rays can enhance or reduce the primary X-ray signal produced by the electron beam.

X-ray fluorescence may be of particular concern in some specimens such as metal foils which contain a certain combination of elements such as Fe and Cr. Iron has a higher excitation potential than chromium, and produces K_α X-rays (6.40 keV) which are at a convenient energy for ionising Cr (5.98 keV) to produce CrK_α X-rays (5.41 keV). In practice this tends to suppress the FeK_α X-ray intensity leaving the specimen and to enhance the

CrK_α line. X-ray fluorescence may also be a problem in the study of dense objects such as mineral particles or precipitates in extraction replicas. Fluorescence of some elements may also be caused by X-ray continuum generated in the specimen (§ 2.4.4). X-ray fluorescence is discussed fully by Wittry (1962) who presents graphs showing the dependence of X-ray fluorescence on accelerating voltage, on the concentration of the exciting element, and on the atomic numbers of the exciting and excited elements.

2.4.3 X-ray absorption

In addition to the X-ray absorption which accompanies the X-ray fluorescence described above, there also exists another type of X-ray absorption. This arises when X-rays are simply stopped en route to the X-ray detector or are scattered out of the line to the detector. This absorption may occur within the specimen or while passing through window materials, diffracting crystals, or even air, after leaving the specimen. X-rays of higher energy (from heavier elements) are less easily absorbed than those from lighter elements with low energies.

If the absorption or scattering occurs within the specimen the emergent X-rays may interact with tightly-bound or loosely-bound electrons and may or may not lose intensity. Obviously such scattering is more likely to occur in thicker specimens and will depend on the direction in which the X-rays leave the specimen. X-rays leaving the specimen at a high angle θ_1 (Fig. 2.4)

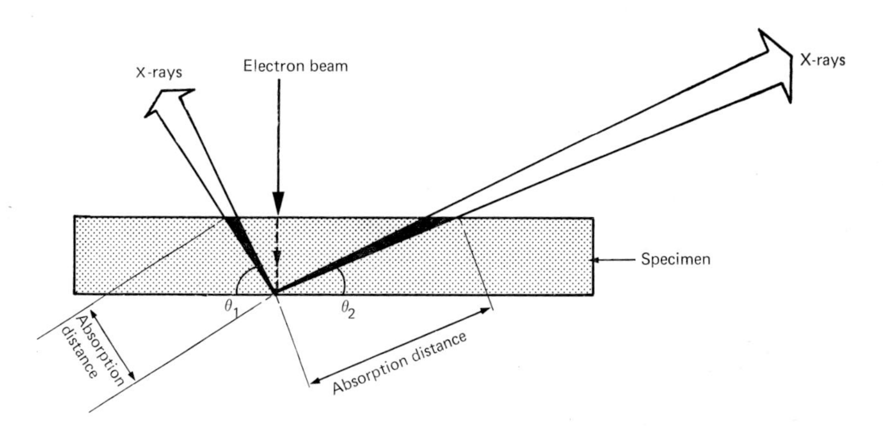

Fig. 2.4. Absorption in a thick specimen for varying X-ray take off angle θ. The more shallow the angle (e.g. θ_2), the greater the absorption, and so detectors are usually positioned at high angles (e.g. θ_1) for collection.

are less likely to be self-absorbed than those leaving at a more grazing incidence θ_2. For this reason, relatively high X-ray 'take off' angles (θ) are employed in most instrumental designs.

Absorption can occur outside the specimen in a number of ways. In many instrumental arrangements, the X-rays must pass through thin windows before reaching the detector (see § 5.2.6d). Absorption in these windows is much more serious than self-absorption in the specimen, which for many thin specimens may be ignored. Different materials absorb X-rays to different degrees and are characterised by their mass absorption coefficients.

The process of absorption is described by the relationship:

$$I_t = I_0 e^{-\mu x}$$

where I_t is the transmitted intensity of the X-ray beam, I_0 is the incident X-ray intensity, μ is the *linear absorption coefficient*, and x is the thickness of the absorbing material. A more convenient way of expressing this is:

$$I_t = I_0 \, e^{-\mu/\rho \, \cdot \, \rho x}$$

where ρ is the density of the absorbing material. The factor μ/ρ is often used in tables of *mass absorption coefficients* and is related to the energy of the X-rays being absorbed. A useful set of such tables is given by Heinrich (1968).

The effect of increasing the thickness of the X-ray windows used in detector systems on the absorption of X-rays of various energies is discussed in § 5.2.6d. The combined effects of absorption and fluorescence, together with backscattering of electrons from the specimen surface and attenuation of electrons in the specimen, constitute an overall correction in the X-ray radiation emitted from thick specimens (known as the *ZAF correction*). This is considered in more detail in Chapter 6.

2.4.4 *X-ray continuum (white radiation)*

The characteristic X-ray radiation described above occurs when primary electrons interact with orbital electrons of the atom. Another type of radiation occurs when the primary electron beam interacts with the nucleus of an atom (Fig. 2.1). The electrons are decelerated in the field surrounding the charged nucleus and are effectively scattered inelastically. The amount of energy lost during this event covers a continuous range from zero up to the initial energy of the primary electrons. Figure 2.5 shows the spectrum of this continuous radiation from a sample composed of 3 elements, Z_1, Z_2 and Z_3. Each of these elements will produce a contribution to the total emitted

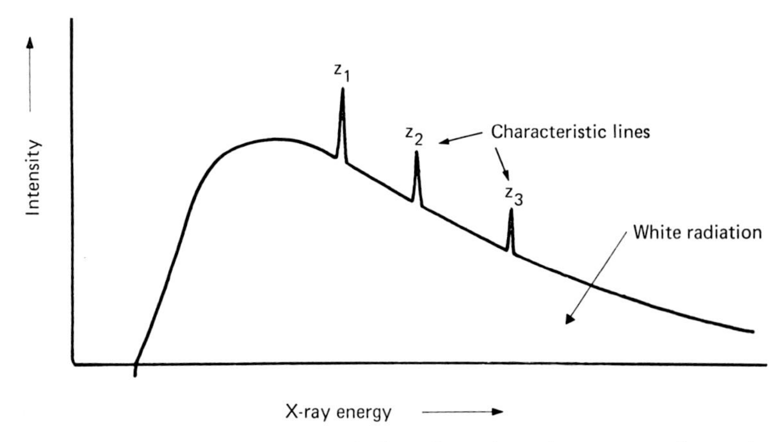

Fig. 2.5. The spectrum of characteristic lines from three elements superimposed upon a background of white radiation.

continuous radiation and each will simultaneously produce characteristic emissions, shown here for simplicity as single spectral lines. This continuous radiation is called '*bremsstrahlung*' '*continuum*' or '*white radiation*'. 'White radiation' is the term most often used and we shall employ it here (Hall 1971).

The white radiation forms the major part of the X-ray background upon which the characteristic lines are superimposed. The amount of white radiation produced by the specimen is a function of the total numbers of atoms, of all kinds, in the specimen, unlike the characteristic radiation which is specific for one particular element. Thus the white radiation provides a means of measuring specimen mass thickness, a necessary procedure when performing quantitative analysis on thin sections. The information available from the white radiation will be considered in detail in the chapter devoted to quantitation (Chapter 6).

2.4.5 *X-ray spectra*

The possible transitions which may occur between electron orbits within an atom were illustrated in Fig. 2.2 while Table 2.2 lists the detectable X-ray emissions due to the transitions which occur in a zinc atom (atomic number 30). This table serves as a useful example of the spectrum produced by an element. Lighter elements produce fewer lines than shown here and heavier atoms have more complex spectra.

In the case of zinc, the most intense lines are the $K_{\alpha1}$ and $K_{\alpha2}$ lines. Next in intensity are the $K_{\beta1}$, $K_{\beta2}$, $K_{\beta3}$ lines etc., and then the $L_{\alpha1}$, $L_{\alpha2}$, $L_{\beta1}$, $L_{\beta2}$ lines

TABLE 2.2

X-ray emissions from zinc after ionisation. The $K_{\alpha 1}$ line is the most easily detected.

X-ray line	*Common notation*	*X-ray energy (keV)*	*Relative intensity*
α_2 KL_{II}	$K_{\alpha 2}$	8.615	Medium high
α_1 KL_{III}	$K_{\alpha 1}$	8.638	High
$\beta_{1,3}$ $KM_{II\ III}$	$K_{\beta 1}$	9.572	Medium low
β_2 $KN_{II\ III}$		9.658	Low
β_5 $KM_{IV\ V}$		9.650	Low
$\beta_{3,4}$ $L_I\ M_{II\ III}$	$L_{\beta 3}$	1.107	Low
η $L_{II}\ M_I$		0.906	Low
β_1 $L_{II}\ M_{IV}$	$L_{\beta 1}$	1.034	Low
l $L_{III}\ M_I$		0.884	Low
$\alpha_{1,2}$ $L_{III}\ M_{IV\ V}$	$L_{\alpha 1}$	1.011	Medium
$M_{II\ III}\ M_{V\ V}$		0.079	Low

and so on. With the heavier elements it is not always possible to excite the K_α lines of the atom. For example, the K line of uranium has an excitation potential of 115.6 keV, and a primary electron beam with an energy of only 80 keV is not able to cause ionisation. Consequently, the L_α line, which has an energy of 13.6 keV, is more often generated. Another consideration is the ability of the X-ray detector to efficiently detect a particular X-ray energy. Using uranium again as an example, it may be easier for the detector to measure the intensity of the M_α line, which occurs at 3.17 keV, than of the L_α line, although the intensity of the L_α line is the greater (see § 5.2.6a).

Another problem is that of the overlapping of spectral lines when a number of elements are present in the specimen and are all producing X-ray spectra. Sometimes it may not be possible to distinguish two particularly close X-ray lines in one part of the energy range and there may be a need to observe another part of the spectral range where the lines are further apart. These problems are considered further in § 3.3. where detector sensitivity and resolution are discussed.

References

Bearden, J. A. (1964), X-ray wavelengths, N.Y. 0-10586, U.S. Atomic Energy Commission, Oak Ridge, Tennessee, U.S.A.

Birks, L. S. (1969), X-ray spectrochemical analysis (Interscience, New York).

Burhop, E. H. S. (1952), The auger effect (Camb. Univ. Press).

Green, M. and V. E. Cosslett (1961), The efficiency of production of characteristic X-radiation in thick targets by a pure element, Proc. Phys. Soc. *78*, 1206.

Hall, T. A. (1971), The microprobe assay of chemical elements, in: Physical techniques in biological research, Vol 1A, p. 393, ed. G. Oster, (Academic Press, New York).
Heinrich, K. F. J. (1968), Quantitative electron probe microanalysis, N.B.S. Technical Note 298, Washington D.C., U.S.A.
Moseley, H. G. J. (1914), The high frequency spectra of the elements, part II, Phil. Mag. *27*, 703.
Wittry, D. B. (1962), Fluorescence by characteristic radiation in electron probe microanalyser, USCEC Rept. 84-204, Univ. of Southern Calif., Los Angeles, U.S.A.

Chapter 3

Collection of X-rays

3.1 *Comparison of thick and thin specimens*

Originally X-ray microanalysis was developed for the study of thick specimens, that is specimens in which the electron beam is totally stopped, and many instruments, such as the electron-probe microanalyser, are still produced and used to analyse this type of sample. Imaging the specimen is achieved either by collection of electrons which have been scattered from the surface of the specimen or by the incorporation of a light-optical arrangement into the microscope. Generally speaking, the requirements of image resolution in thick specimens are not particularly severe and greater emphasis is placed on the X-ray detection capabilities of such instruments.

3.1.1 *Spatial resolution*

The difficulties of achieving adequate resolution in a thick sample, and the advantages of using thin specimens are illustrated in Figs. 3.1 and 3.2 respectively. An electron beam incident upon the surface of a very thick specimen will diffuse several micrometres into the sample (depending on its composition) before all of the electrons are stopped (Fig. 3.1). As these electrons travel through the specimen they ionise atoms, provided the electron energy is greater than the critical ionisation potential E_c. As soon as this energy falls below E_c no further ionisations occur but the electrons continue decelerating until they have zero energy and still produce a certain amount of white radiation. Characteristic X-rays are thus seen to come from a pear-shaped volume of the specimen below the surface and the depth of this volume depends on the penetrating power of the electron beam, that is

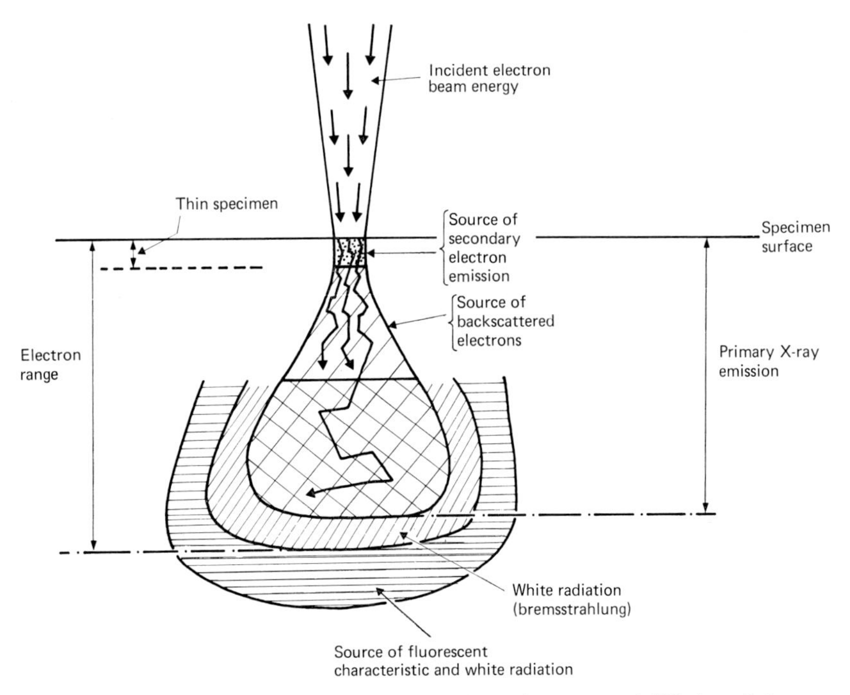

Fig. 3.1. The penetration of electrons into a bulk specimen. Lateral diffusion of electrons occurs and electrons generate characteristic X-rays until their energy falls below the critical ionisation potential E_c. They are brought to rest at $E = 0$. White radiation may generate fluorescent X-rays in a volume beyond the trajectory of the primary electron beam. (Courtesy of John Wiley and Sons Inc., New York.)

on the primary electron energy (accelerating voltage), and on the specimen composition. Spatial resolution of analysis is therefore a function of these two parameters.

With an ultrathin specimen less than 200 nm thick, such as those examined in transmission electron microscopes (TEM), such difficulties are virtually non-existent. The electron beam diffuses laterally to a very small extent within the first 100 nm of the specimen (Fig. 3.2). Russ (1972) has calculated that for a 50 kV electron beam 90% of the X-ray emission comes from a cone having a solid angle approximately 30° to the vertical. This means that for a biological specimen 100 nm thick an electron probe of diameter 100 nm at the upper surface will diffuse laterally to a width of < 200 nm by the time it reaches the lower surface. For many purposes this diffusion can be considered to be negligible. However, when very small probe diameters are being used, such a spread may prove embarrassing. At accelerating

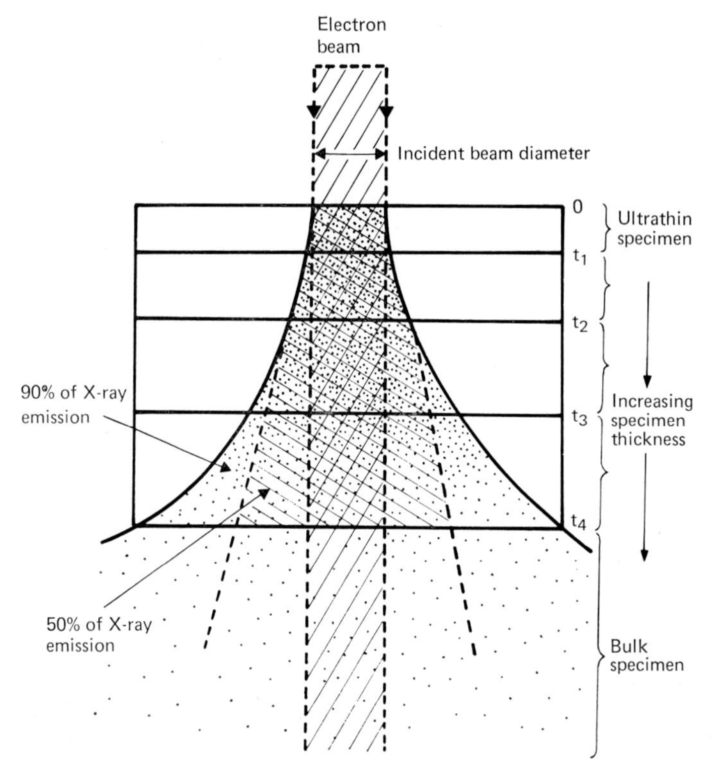

Fig. 3.2. Lateral diffusion of electrons entering the surface of a specimen. In an ultrathin specimen the diffusion is negligible. (Courtesy of John Wiley and Sons Inc., New York.)

voltages above 50 kV the lateral diffusion within an ultrathin specimen is reduced and is usually ignored. Thus the X-rays in ultrathin specimens are assumed to come from the exact illuminated area seen on the fluorescent screen (Fig. 3.3). For thicker specimens, however, penetration and lateral diffusion increase with accelerating voltage (Hall 1971).

3.1.2 *Absorption and fluorescence*

The processes of X-ray fluorescence and absorption, described in § 2.4.2 and § 2.4.3 above, are particularly important problems in the study of thick specimens (§ 6.1). In ultrathin sections, however, the path through which the X-rays have to pass to emerge from the specimen surface is very short and for most purposes these effects may be ignored.

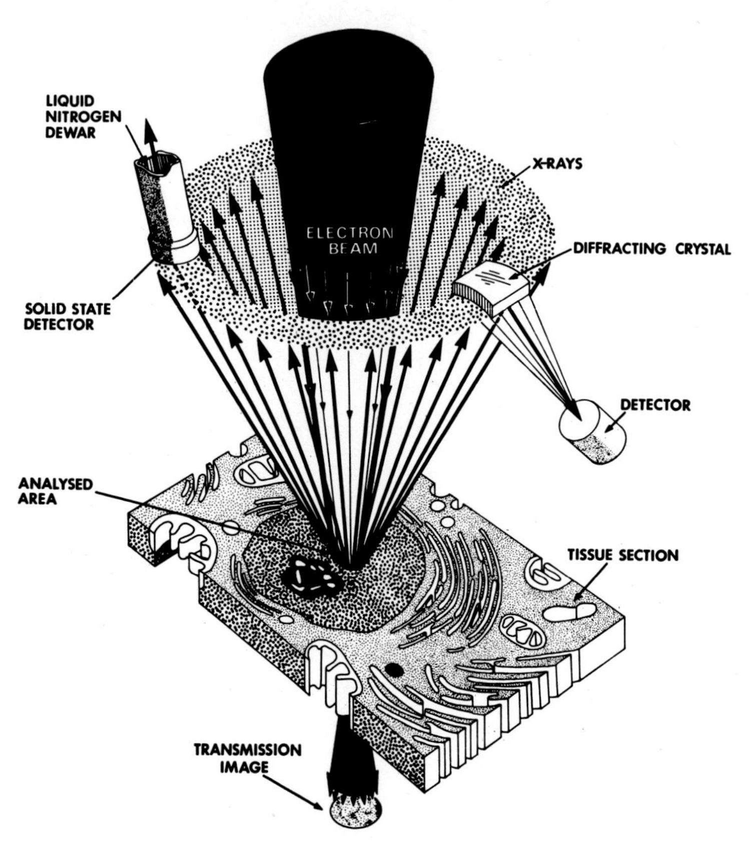

Fig. 3.3. X-ray analysis from an ultrathin specimen. The region of analysis is accurately defined by the electron probe diameter. (Courtesy of John Wiley and Sons Inc., New York.)

3.1.3 Image resolution

Since the image of a thin specimen is always in view on the fluorescent screen (Fig. 3.3), the area chosen for analysis can be located very accurately. An even greater advantage of thin specimens is that the resolution of the transmission image, which is about 10–20 times greater than that of a scanning electron microscope (SEM) image of a bulk specimen, allows the analysis to be correlated with fine details of ultrastructure. In most electron-optical arrangements (§ 3.4) the X-ray analysis can be performed without seriously impairing the normal transmission imaging facilities of the microscope.

3.1.4 Sensitivity of analysis

As will be discussed in some detail later (Chapter 6), the sensitivity obtainable in the analysis of thin specimens is enhanced by the fact that the elements being detected are present within a very limited volume. Whereas in a thick specimen X-ray emission comes from a large volume and may consequently contain a significant background signal (mostly from white radiation), in thin specimens this background is much reduced and small concentrations of elements are hence more easily detectable. However, because of the 'thinness' of transmission specimens the amount of material actually available for analysis is very small and hence the total X-ray intensity emitted from the specimen region being irradiated is also very small. A great deal of attention must therefore be paid to instrumental and operational factors affecting X-ray collection and suppression of background.

The various factors discussed in this chapter determine to a great extent what type of specimen is best suited for a particular application and which is the best type of instrument to use (see § 8.1). There are a number of different electron-optical arrangements for the analysis of thick and thin specimens and these will be described in § 3.4, but first the different types of X-ray detector that can be incorporated into these microscopes will be considered.

3.2 X-ray detectors

The purpose of the X-ray detector is to receive as many as possible of the X-rays that emerge from the area or volume of the specimen being bombarded by electrons, and to analyse the various energies (or wavelengths) of these X-rays in order to identify the elements in the source. There are, however, restrictions on the design of the detector, since to have a high collecting power it must be placed as close as possible to the specimen and yet not interfere with the electron optics of the microscope.

There are basically 3 types of X-ray detector in use with electron-optical systems: the *wavelength dispersive crystal spectrometer*; the *energy dispersive solid state detector*, and the *gas-flow proportional counter*. Each detector has certain advantages and each will be considered in detail and its particular advantages and disadvantages enumerated before describing how it may be combined in an electron-optical column. Investigators are often confused by the superlative claims made by manufacturers of X-ray detectors. It is hoped that the following descriptions will help to disperse some of the confusion and assist in making the correct choice of a particular detector for a particular problem.

3.2.1 Wavelength dispersive crystal spectrometers

The principle of a wavelength dispersive spectrometer is illustrated in Fig. 3.4. Under electron bombardment the various elements in the specimen emit X-rays with a range of wavelengths ($\Sigma\lambda$). These X-rays leave the specimen over the whole solid angle but because of the finite size of the crystal (generally about 2.5 × 1 cm) only a narrow cone of X-rays is accepted by the spectrometer. The curved crystal reflects a fraction of these X-rays into a suitable detector. The fraction of the X-ray beam which is reflected depends on the principle of '*diffraction*' – only one particular wavelength (λ) is strongly reflected at a certain angle to the exclusion of all others.

This principle is described by Bragg's law which states that

$$n\lambda = 2d \sin \theta$$

where n is an integer, λ is the wavelength of the X-rays which are precisely diffracted, d is the lattice spacing of the particular planes in the crystal giving rise to the diffracted beam, and θ is the angle of incidence (and of reflection) of the X-rays arriving at the crystal (Fig. 3.4).

Thus for a crystal of known lattice spacing, d, and for a selected angle of

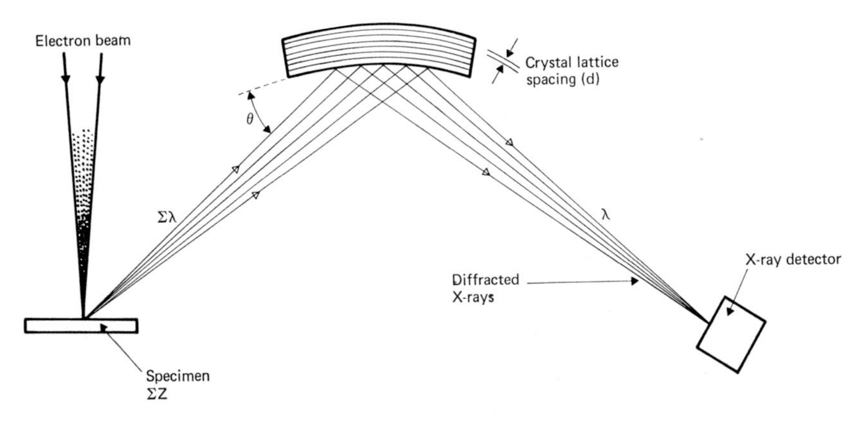

Fig. 3.4. Collection of X-rays by a wavelength dispersive crystal spectrometer. X-rays are generated having a range of wavelengths, $\Sigma\lambda$, but only one wavelength, λ, is selectively diffracted to the detector corresponding to the angle, θ, at the diffracting crystal. The crystal has a lattice spacing, d, for diffraction. (Courtesy of John Wiley and Sons Inc., New York.)

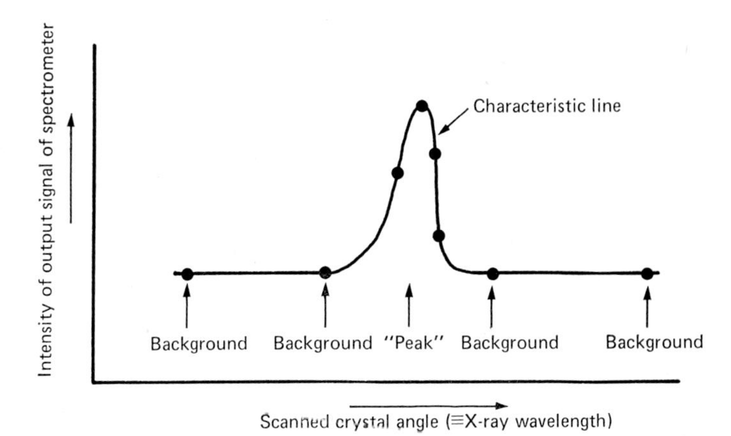

Fig. 3.5. Scanning the crystal of a wavelength dispersive spectrometer through its angular range produces a peak of intensity when the characteristic wavelength is reached.

incidence, θ, the wavelength of the diffracted X-rays entering the detector can be calculated. In practice the crystal is rotated through a range of angles until a '*peak*', or maximum of intensity, is noted in the detector (Fig. 3.5) at which point the conditions of Bragg's law are satisfied and the wavelength of the X-rays detected can be identified.

3.2.1a Diffracting crystal

The crystal itself is bent or ground to a radius R in order to bring the diverging beam of X-rays emitted by the specimen to a focus at different points on a circle of radius R/2, the *Rowland circle* (Fig. 3.6a). The detector is placed on this circle and is free to move as the angle of the crystal is changed relative to the direction of the X-ray beam.

Two types of crystal spectrometer geometry are used. In one, called the *Johann geometry*, the crystal is bent to a radius R (Fig. 3.6a), while in another a more precise arrangement, called the *Johansson geometry*, is used (Fig. 3.6b). In this the crystal is first bent to a radius R and then ground to a radius R/2 so that all points of focus lie very exactly on a circle. These latter crystals are quite difficult to make and are rather expensive so that the Johann geometry is the one most commonly encountered.

The range of wavelengths that such a crystal can focus depends upon the range of angles, θ, that it can be rotated through. Consequently a crystal can only cover a range of a few tenths of a nanometre (1 nm = 10 Angstrom

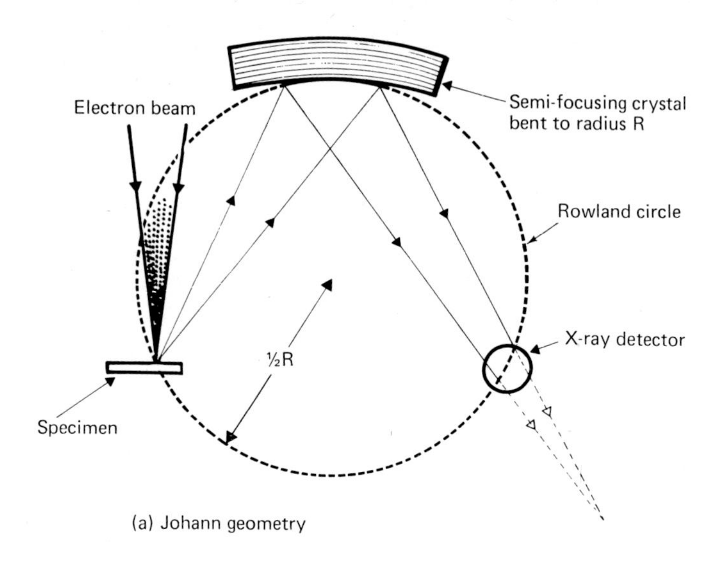

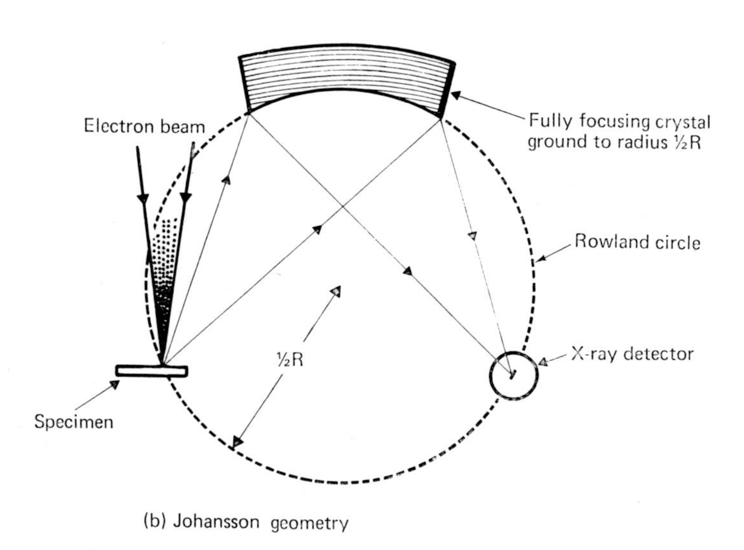

Fig. 3.6. X-ray focusing in crystal spectrometers (a) Johann and (b) Johansson systems. The specimen, crystal and detector are maintained on the circumference of the Rowland circle for all tuned positions. (Courtesy of John Wiley and Sons Inc., New York.)

TABLE 3.1

Some commonly used crystals in wavelength dispersive spectrometers

Crystal	*Reflecting planes*	*Wavelength range (nm)*	*Detectable atomic number range for K_{α} radiation*
Lithium fluoride (LiF)	200	0.1 –0.38	19–35
Ortho-phthalate potassium hydrogen (KAP)	001	0.45–2.5	11–14
Ortho-phthalate rubidium hydrogen (RbAP)	001	0.2 –1.8	11–14
Gypsum	020	0.26–1.5	11–14
Ammonium dihydrogen phosphate (ADP)	101	0.18–1.03	12–21
Ethylene diamine tartrate (EDT)	020	0.14–0.83	14–22
Pentaerythritol (PET)	001	0.14–0.83	14–22
Germanium (Ge)	111	0.11–0.60	16–34
Sodium chloride (NaCl)	200	0.09–0.53	16–37

units) of wavelength, corresponding to the X-ray lines of only a few elements in the periodic table. To extend this range, a diffracting spectrometer is often equipped with a number of crystals having different *d* values or lattice spacings, so that for the same range of angles (θ), the wavelength (λ) range is extended and a greater number of elements covered.

The types of crystals commonly used in diffracting spectrometers, and the ranges of elements which they can cover are listed in Table 3.1. Crystals may be thin slices of naturally occurring minerals, such as mica or gypsum, or may be synthetically prepared, such as stearate crystals which are multiple monomolecular layers of soap, and which are used for the diffraction of the X-rays of very long wavelengths from light elements. Gratings with ruled surfaces are also used in a Johann geometry for the detection of long wavelength X-rays. For good resolution and highly efficient diffraction of the X-ray lines, the crystal must be of excellent quality with no flaws or blemishes. The mechanism that governs the position of the crystal relative to the specimen and to the detector must be of the highest accuracy to achieve reproducibility.

3.2.1b Spectrometer detector

After being selectively diffracted by the crystal the X-rays of a particular wavelength pass through a collimator (§ 5.2.6e) towards the spectrometer

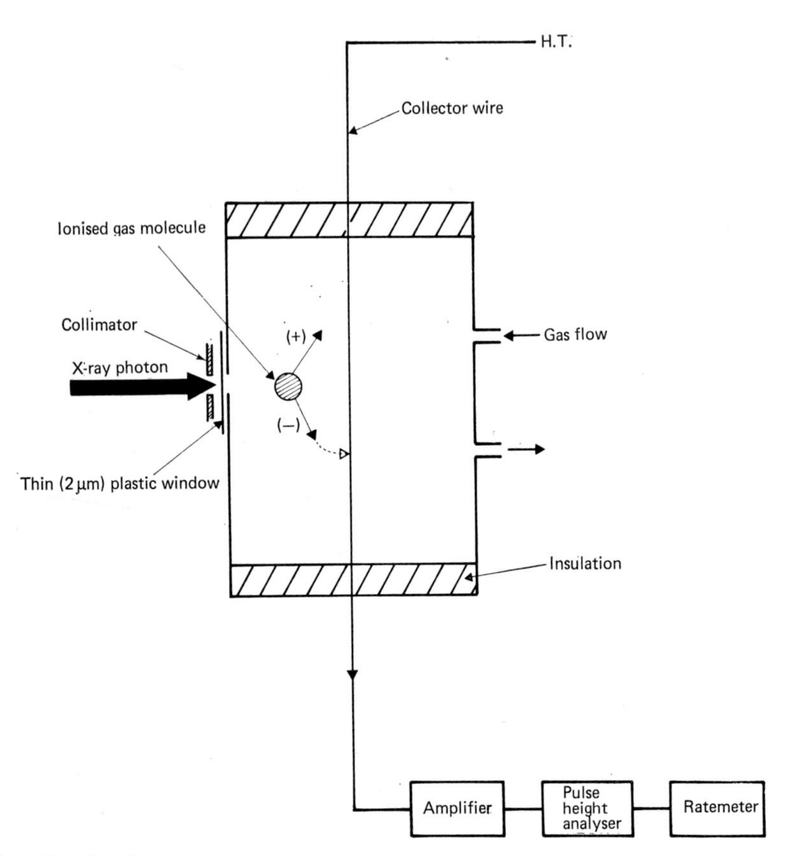

Fig. 3.7. Gas-flow proportional counter (GFPC) used either alone or in conjunction with a crystal spectrometer. The X-rays are collimated via a slit and enter the detector through a thin window. A positive pressure of gas is maintained inside.

detector. The most commonly used type of detector is a proportional counter. This consists of a gas-filled cylinder (Fig. 3.7) with a wire along its axis. The wire is insulated from the cylinder, and is at a positive potential (generally around 1500 V DC) with respect to the detector body which is earthed. The X-rays enter the detector through a thin plastic window, which is typically a 2 μm thick film of polycarbonate or polypropylene, and which is required to maintain the positive pressure of gas in the cylinder.

Proportional counters may be of a sealed or gas-flow type. In sealed counters relatively thick X-ray windows are used to prevent gas leakage and the gas is permanently sealed in. In the gas-flow type ultrathin windows (sometimes down to 100 nm) are used and the gas has to be continuously pumped through the cylinder to maintain a constant flow rate. The gas-flow type is also used when very weak X-rays (long wavelengths) are to be detected and where thicker windows would cause attenuation. The gas often used for proportional counters is a mixture of argon and methane, and is supplied from a free standing cylinder at the side of the microscope.

When an X-ray photon enters the counter it collides with a gas molecule giving up a portion of its energy and ionising the molecule. This produces an electron-ion pair. The positive potential of the wire attracts the electron which then gains enough energy to ionise other molecules. The net result is an avalanche of electrons travelling to the wire to produce an electrical pulse with an amplitude which is dependent on the energy of the original X-ray photon. The X-rays are thus converted to electrical signals and may be analysed and processed by suitable *nucleonics* (§ 5.2.6f). The intensity of the original X-ray beam, which is related to the number of atoms of the element being analysed (§ 2.4.1) is thus translated into a proportional number of electrical pulses.

3.2.1c Operation of the spectrometer

In the Johann and Johansson types of spectrometer the specimen, crystal and detector are always maintained on the circumference of the Rowland circle (Fig. 3.6). A common design is termed a linear focusing arrangement in which the crystal changes angle while moving along a linear track away from the specimen. The mechanism for the spectrometer adjustment must be very precise, since a slight movement of the crystal away from the exact diffracting angle produces a rapid fall in detected intensity. This linear arrangement ensures that a very good peak-to-background ratio is achieved, but some difficulties in reproducibility may be experienced (§ 5.3.3a).

Modern crystal spectrometers can be evacuated and either connected to the microscope vacuum system directly or separated by a thin (0.1–10 μm) plastic window such that absorption of long wavelength X-rays by air does not occur. If windows are included then long wavelength (low energy) X-rays and stray electrons may be filtered out.

3.2.2 Gas-flow proportional counter

As well as being used as a detector within the crystal spectrometer, the gas-flow proportional counter can be used in its own right to detect X-rays directly without any other spectrometer device. The basic design is as described above, except that the detector is usually larger to accept a greater solid angle of X-rays from the specimen. Thus, unlike the crystal spectrometer where only one X-ray wavelength enters the detector, the gas-flow proportional counter receives X-rays of all wavelengths from the specimen simultaneously. Since each X-ray produces a unique pulse amplitude, i.e. a voltage pulse inversely proportional to the X-ray wavelength (or proportional to its energy), a range of X-ray wavelengths from a range of elements will produce a range of pulse amplitudes in the detector. These are all displayed on a *cathode-ray oscilloscope* (CRO) (§ 5.3.5).

The '*dead-time*' of such a detector is the time that it takes for an X-ray photon to register as a pulse and for the pulse to decay. During this time the detector is effectively dead, or unable to detect other incoming pulses. A typical dead-time is one microsecond, permitting count rates of up to 10,000 counts/sec. However, the eye is unable to differentiate pulses in this short time interval and a CRO effectively provides an integrated spectrum of pulses that appear to be all recorded simultaneously. In order to differentiate pulses of different amplitudes, and therefore corresponding to different elements, facilities for *pulse height analysis* (PHA) are necessary. In this arrangement only the pulses of a certain height are allowed to pass through a '*voltage window*' to be finally counted (5.2.6f).

Even though each X-ray photon from a given atom has exactly the same energy, the gas ionisation process in the detector is statistical in nature. This results, in practice, in pulses with a short range of amplitudes being produced for the same X-ray energy. Thus proportional counters cannot resolve elements closer together than two or three atomic numbers in the periodic system. Such a detector is useful, however, if used in conjunction with the wavelength crystal spectrometer (§ 3.2.1) which cannot display a whole spectrum of X-ray energies simultaneously.

Another important use for a gas-flow proportional counter is to collect the 'white radiation' (§ 2.4.4) for the measurement of specimen '*mass thickness*' in thin sections as discussed in full later (§ 5.3.5b).

3.2.3 Energy dispersive analyser

As discussed above, a limitation of the crystal spectrometer is that it is unable, by definition, to detect and display simultaneously all X-ray energies leaving the specimen. The gas-flow proportional counter is able to do this but with very limited resolution.

A semi-conductor device which has become available in the last few years

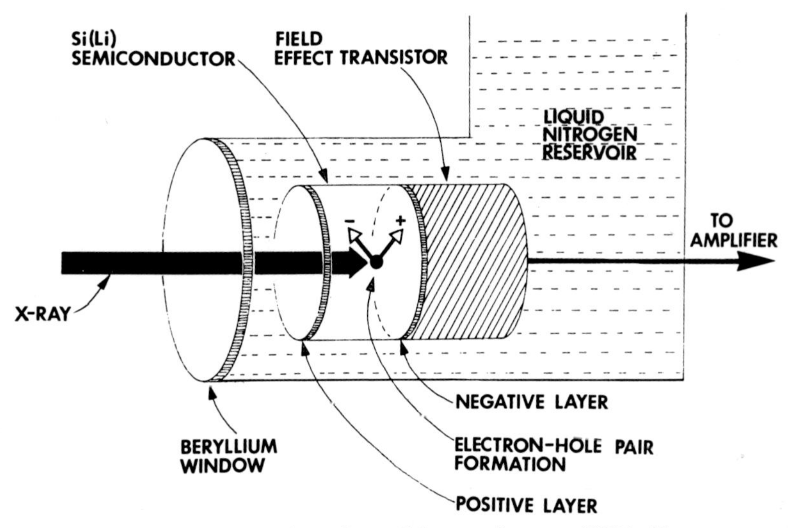

Fig. 3.8. Schematic representation of a solid state detector (SSD). X-rays enter the detector through a thin beryllium window and produce electron-hole pairs within the semi-conductor crystal. The ionisations produce pulses which are amplified by a field effect transistor (FET) built into the rear of the crystal. (Courtesy of John Wiley and Sons Inc., New York.)

and which is rapidly gaining in importance is the *solid state detector*. This consists of a silicon radiation detector with a surface area in the range 5 to 200 mm^2 located between two metal electrodes across which a bias voltage is applied (Fig. 3.8). A layer of lithium is partly diffused into the silicon crystal forming a semi-conductor, Si(Li). The detector crystal is kept under high vacuum and at liquid nitrogen temperature. X-rays entering the detector through a thin beryllium window create electron-hole pairs by ionisation within the detector. The number of these pairs is proportional to the X-ray energy and is equal to the ratio of the X-ray energy and the energy required to create a pair. In a Si(Li) detector the energy required to create a pair is about 3.8 eV (electron volts). The total charge produced by a single X-ray photon is integrated by collection at the electrodes and is then fed to a *field effect transistor* (FET) in a pre-amplifier which is positioned close to the detector to reduce noise. The output voltage pulse is then further amplified and passed into a *multichannel analyser* (MCA) in which the pulses are separated in terms of amplitude and stored in memory channels corresponding to these amplitudes. The resulting energy spectrum can be displayed on a CRO which is built into the MCA, or on an XY chart recorder (Fig. 3.9). The spectrum can be recorded on punched paper tape, or teletype, or

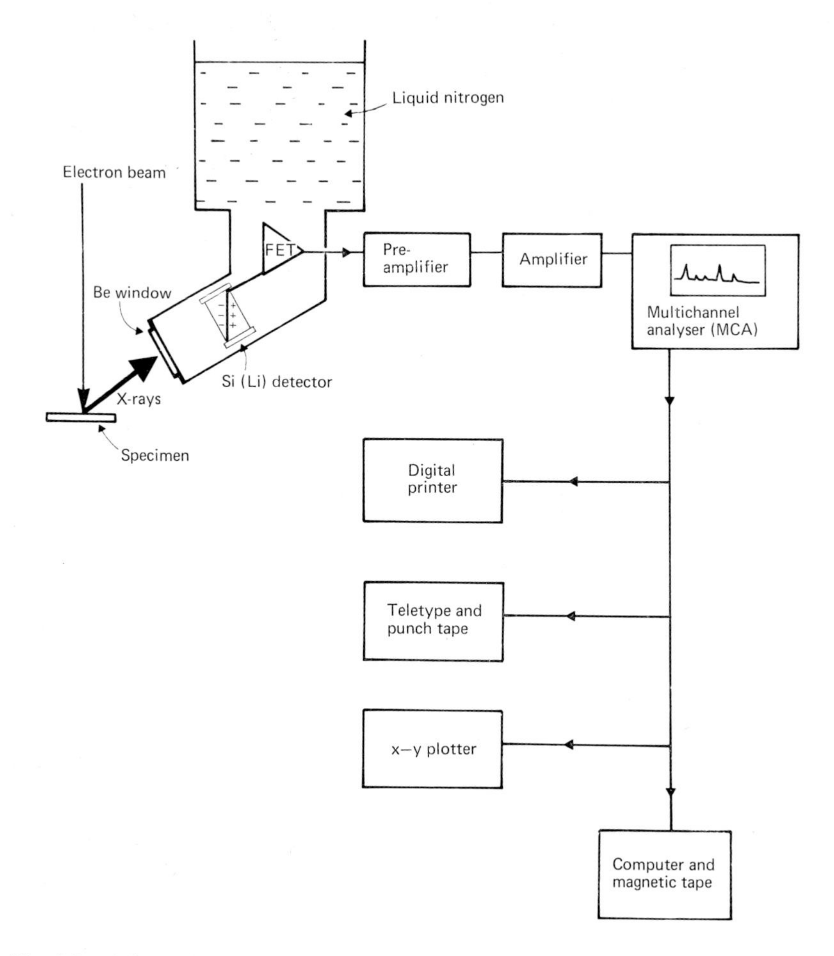

Fig. 3.9. Schematic diagram of the nucleonics associated with an SSD. (Courtesy of John Wiley and Sons Inc., New York.)

transferred directly to a computer or magnetic tape. Computer programmes are available which rapidly read the energy spectrum, apply appropriate correction factors and calculate numerical values for elemental concentrations, etc.

A very great advantage of the energy dispersive analyser, or solid state detector (SSD), is that it can be placed very close to the source of X-rays and so accept a wide solid angle of radiation (Fig. 3.10). This increases detection sensitivity enormously and thus provides better statistical data.

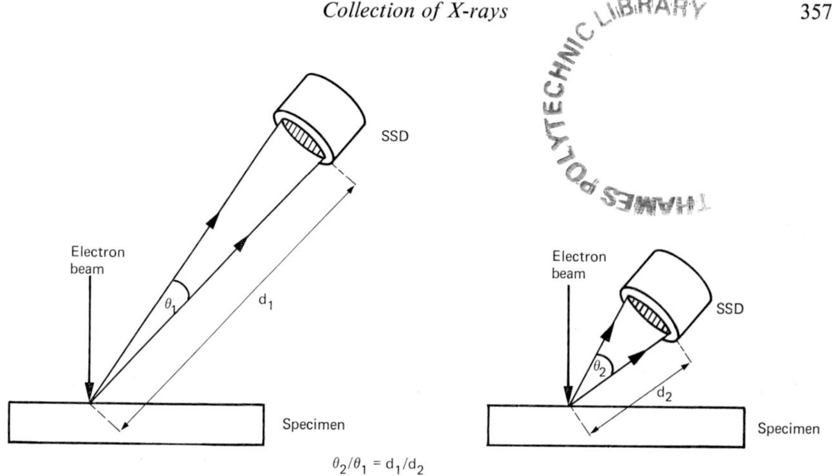

Fig. 3.10. Collection of X-rays by a solid state detector (SSD). Moving the detector closer to the specimen (from d_1 to d_2) increases the solid angle of collection (from θ_1 to θ_2) and hence increases the count rate sensitivity.

3.3 Comparison of X-ray detectors

The type of system chosen for a particular application depends on the nature of the problem. For example, when dealing with biological materials, except teeth or bone, the concentrations of elements are frequently very low ($< 1\%$ by weight) and special conditions govern the choice of the analysing detector and of the electron-optical system. The important features of detecting systems will be considered in turn.

3.3.1 Energy resolution

The energy resolution of a detector may be defined as its ability to separate two adjacent peaks in the energy spectrum. The resolution is generally measured as the *full width* of the peak at *half* the *maximum* intensity *(fwhm)* (Fig. 3.11), and is usually quoted at the energy position 5.9 keV in the spectrum. Alternatively the full width at one tenth maximum is taken as a criterion.

When a number of elements are present together in the specimen region being analysed, there are many other overlapping X-ray lines in the energy spectrum in addition to the major X-ray lines. For example the K_α line of phosphorus occurs at 2.01 keV whereas the M_α line of osmium, a frequently employed fixative and stain in electron microscopy, occurs at 1.91 keV, only 100 eV away. Such overlaps occur frequently in biological studies, especially

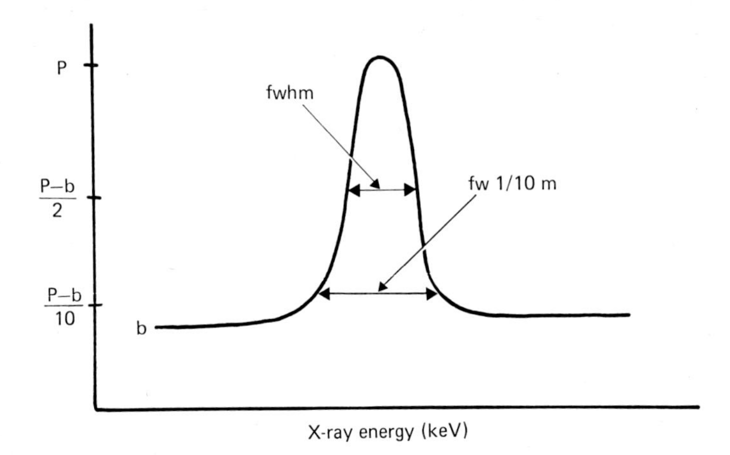

Fig. 3.11. Resolution of characteristic X-ray lines. The full width half maximum (fwhm) position is generally taken as the resolving power of a detector, but often the full width at one tenth maximum is quoted.

TABLE 3.2

Some major X-ray lines in the 0–10 keV range

Element	*Characteristic line*	*eV*	*Element*	*Characteristic line*	*eV*
Cu	L_α	930	Sn	L_α	3443
Zn	L_α	1011	Sb	L_α	3604
Na	K_α	1041	Ca	K_α	3691
Mg	K_α	1253	Sb	L_β	3843
Al	K_α	1486	Ti	K_α	4510
Si	K_α	1740	Cr	K_α	5414
Si	K_β	1835	Mn	K_α	5898
Os	M_α	1910	Fe	K_α	6403
P	K_α	2013	Ni	K_α	7478
Au	M_α	2123	Cu	K_α	8047
P	K_β	2139	Zn	K_α	8638
S	K_α	2307	Os	$L_{\alpha 2}$	8841
Pb	M_α	2345	Cu	$K_{\beta 2}$	8905
Cl	K_α	2622	Os	$L_{\alpha 1}$	8911
Ag	L_α	2984	Cu	$K_{\beta 1}$	8977
U	M_α	3170	Zn	K_β	9572
K	K_α	3314	Au	$L_{\alpha 2}$	9628
U	M_β	3336	Au	$L_{\alpha 1}$	9713

when heavy metals are employed to provide contrast, or are used in cytochemical techniques. Table 3.2 lists the major X-ray lines obtained from each element and shows how close many of these lines can be. The ability to resolve two such lines depends on the relative peak heights, that is on the relative concentrations of the elements. Heavy metal stains are frequently present in thin biological tissue sections in concentrations much greater than the naturally occurring elements and resolution requirements vary.

The gas-flow proportional counter has an energy resolution of 1000 eV. Since the major (K_α) lines of the elemental spectrum are separated by only about 200 eV at low atomic numbers ($Z = 11$), and by about 500 eV at 5.9 keV (K_α of manganese, $Z = 25$), this type of detector is useful only for giving an approximate idea of the positions of the peaks in the spectrum.

The solid state detector currently commercially available has a resolution of about 150 eV at 5.9 keV and this resolution is being steadily improved. The performance can also be improved by computer techniques of data smoothing by which the noise and statistical fluctuation are smoothed out, and by *deconvolution* (§ 5.3.4g) by which individual peaks can be removed from a complex spectrum. Physical limitations (thermal and quantum noise) probably place the limit on the resolution likely to be achieved with a solid state detector at about 100 eV. To obtain maximum resolution and minimum noise the detector is maintained at liquid nitrogen temperature. The resolution available increases with decreasing energy in the spectrum. Thus, 150 eV resolution may be possible at 5.9 keV, while at 1.04 keV (sodium) it could be as low as 120 eV.

Crystal spectrometers function at a much better resolution by the nature of their operation. The technique of wavelength dispersion by diffraction ensures that only those X-rays of a very precise wavelength (or energy) reach the detector. The limitations to this system lie in the shape and quality of the diffracting crystal, in the X-ray detector (gas counter), and in the electronics (noise). Resolutions better than 10 eV can be achieved with such a system, depending on the crystal chosen for analysis.

3.3.2 Sensitivity

The ability to resolve X-ray lines is insufficient unless the detector system also allows adequate X-ray counts to be accumulated in reasonably short analysis times. However, a distinction must be made here between 'count rate sensitivity' and 'mass detection sensitivity'.

The solid state detector has the great advantage of being able to be placed very close to the specimen (sometimes just a few millimetres away). This greatly increases the solid angle of collection of X-rays and so raises the count rate. Thus a greater number of X-ray photons can be detected in the same time or the same number in a shorter time. This means that a lower electron signal can be employed to generate the same X-ray signal, thus allowing a smaller probe diameter and reducing specimen damage. However, a faster counting rate does not necessarily in itself increase mass detection sensitivity since this depends critically on the ratio of peak counts to background counts.

If *sensitivity* alone is the important factor (i.e. disregarding resolution requirements) then the advantage of being able to move the SSD close to the specimen makes this type of detector very attractive. However this is not possible in all instrumental arrangements, as will be seen later (§ 3.4).

When a thick specimen is being analysed, a great deal of unwanted X-ray radiation is being produced from the specimen by the electron beam. Since a solid state detector is non-discriminating in its acceptance of incoming radiation, this high total intensity may tend to swamp the detector and produce a high 'dead-time', a condition in which the detector is made insensitive to subsequent X-rays (see § 5.3.4a). Thus the advantages of gaining a greater solid angle of X-ray collection by reducing the specimen-to-detector distance may be lost by the effect of dead-time on the detector.

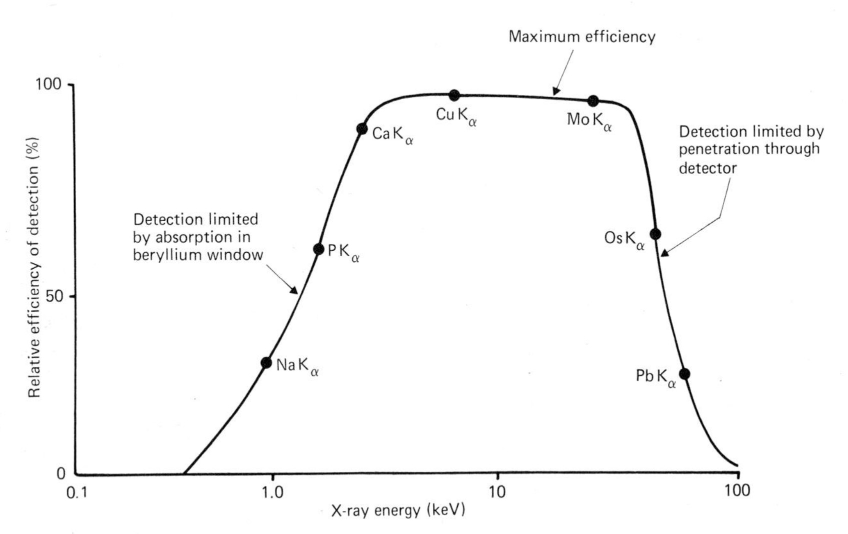

Fig. 3.12. Relative detection efficiency of a typical SSD over the X-ray energy range. (Courtesy of John Wiley and Sons, New York.)

This is particularly true of thick specimens where a lot more of the specimen than the element of interest is producing X-ray emission. With thin specimens, however, there is less background radiation from the matrix and much of the signal is due to the elements of interest in the specimen. Consequently a solid state detector, provided it can be positioned close to the specimen without interfering with the electron optics, is sometimes more sensitive (in terms of count rate) than a crystal spectrometer.

Figure 3.12 demonstrates the efficiency of detection of K_α radiation over the energy range with a typical SSD. The relative detection efficiency is the relative efficiency with which X-rays of a given intensity but of varying energy, that is from different elements, can be detected. This detection efficiency does not include the relative efficiency of X-ray production from each element which depends on other factors such as operating conditions during analysis (§ 5.2.1). The sensitivity falls off at low energy due to absorption in the protective window in front of the detector. High energy X-rays pass right through the detector and are not detected.

With respect to *mass detection sensitivity* (the amount in grams of material present) at the detectable limit of the technique, again it is not possible to generalise because of the variation that can occur from one specimen to the next in such important effects as electron scatter, white radiation, overlapping X-ray lines etc. Sensitivity is discussed further in Chapter 6.

3.3.3 Ease of operation

There are many factors involved in the use of X-ray detectors which require special attention for accurate and sensitive results to be achieved. At first glance it appears that the solid state detector, with its static design and simultaneous display of all the elemental peaks in a single spectrum, is far easier to operate than the relatively laborious method of tuning the crystal spectrometer to different wavelengths for individual elements. What really matters is the type of information which is produced and the reliability of the quantitative results obtained. It may be easier to achieve a qualitative assessment of the specimen composition with the energy spectrum of the SSD, but for a quantitative assessment of a limited number of elements within a specimen the linear response of the crystal spectrometer may be preferred, depending on the type of specimen.

With some of the sophisticated computer programmes available for reading the energy spectrum of the SSD output and converting it to numerical figures, even as weight percentages, the energy dispersive analyser is indeed

TABLE 3.3

Relative advantages and disadvantages of solid state detectors and crystal spectrometers

Solid state detector	*Crystal spectrometer*
Advantages	*Advantages*
1. No X-ray focusing required	1. High resolving power separation of X-ray lines easy
2. High sensitivity high solid angle possible up to 100% detector efficiency	2. Quantitation – small dead time effect X-ray count proportional to elemental content. Best for trace elements
3. No diffraction interference from higher order lines	3. Good peak-to-background ratio gives high sensitivity for trace elements
4. Simple mechanical design no moving parts easily added to electron microscopes	4. Good sensitivity to light elements with suitable crystals
5. Output compatible with computer	
6. High count rate detectability allows smaller probe diameter reduces specimen damage	
7. Whole elemental spectrum displayed rapid qualitative analysis no error from specimen or instrumental changes	
Disadvantages	*Disadvantages*
1. Operative at cryogenic temps large cryostat liquid nitrogen supply necessary	1. Mechanical system errors possible between measurements
2. Occasional compromise necessary in electron-optical system	2. Only one element analysed at a time
3. Inferior energy resolution to crystal spectrometers	3. Possible interference from high order diffraction lines
4. Isolation of detector Be window absorbs low energies lower sensitivity for light elements	4. Peak and background must be measured separately
5. Non-discriminating to X-ray sources stray signals from remote areas high background	
6. Quantitation – poor accuracy at very low concentrations	

a convenient system that is relatively easy to operate once the programming routines have been mastered and the various correction procedures to be applied to the spectrum so produced are understood. In other words, the real difficulty lies in processing the data electronically and interpreting it.

With the crystal spectrometer, however, the problems are mainly mechanical. Difficulties often arise in checking the mechanical alignment of such a system and in the choice of suitable positions in the spectrum for measuring peak and background levels.

The relative advantages and disadvantages of the two types of detector (Table 3.3) will become clearer in Chapter 5, where the operation of the systems is described.

3.4 Electron-optical systems

There are a number of different methods of combining electron microscopy and X-ray microanalysis, each one being more or less suited to a particular problem. The two important parameters to be married together are the best resolution for imaging the specimen and the highest sensitivity available from the X-ray detector system. The two principle alternative electron-optical arrangements are *scanning reflection* and *transmission microscopy* (Agar et al. 1974). The former is employed mainly for thick specimens and the latter for conventional ultrathin specimens. As will be shown later in this chapter, the same X-ray detectors can be fitted to both systems. This book is concerned only with those specimens which are suitable for electron microscope analysis, that is relatively thin specimens as compared with bulk samples. Therefore, attention will be focused on the electron-optical configurations used for analysis of such specimens. A wealth of other publications (see list of further reading in Chapter 1) describe the analysis of bulk material, mainly in the electron-probe microanalyser. This instrument, however, is also useful for the examination of thin samples ($< 10\ \mu m$) and so a description of it will be included here.

3.4.1 Electron-probe microanalyser

The electron-probe microanalyser (EPMA) was the forerunner of all present day systems combining electron microscopy and X-ray detection, and it is still in widespread use today. A typical system is shown schematically in Fig. 3.13. The EPMA relies on a high quality light-optical system for visualisation of the specimen surface and is able to provide only medium resolution in an electron scanning image. The emphasis in design is placed upon the quality of the X-ray detection system. Both wavelength dispersive crystal spectrometers and solid state detectors may be incorporated. The specimen area being analysed is viewed either during, or subsequent to, the analysis using the light-optical system. A backscattered electron scanning image is also available and provides information on the specimen surface topography. The electron beam may be focused to a sub-micron area on the specimen and magnetic coils or electrostatic plates are used to deflect the beam in a scanning raster or to make fine adjustments of the static probe position. The specimen may also be traversed mechanically relative to the electron beam.

Three types of electron signal are available from the specimen when it is bombarded by the primary beam; '*reflected*' electrons scattered from the specimen surface; low energy '*secondary*' electrons which emerge from just below the specimen surface; and the '*specimen current*' which is the net total electron beam absorbed by the specimen.

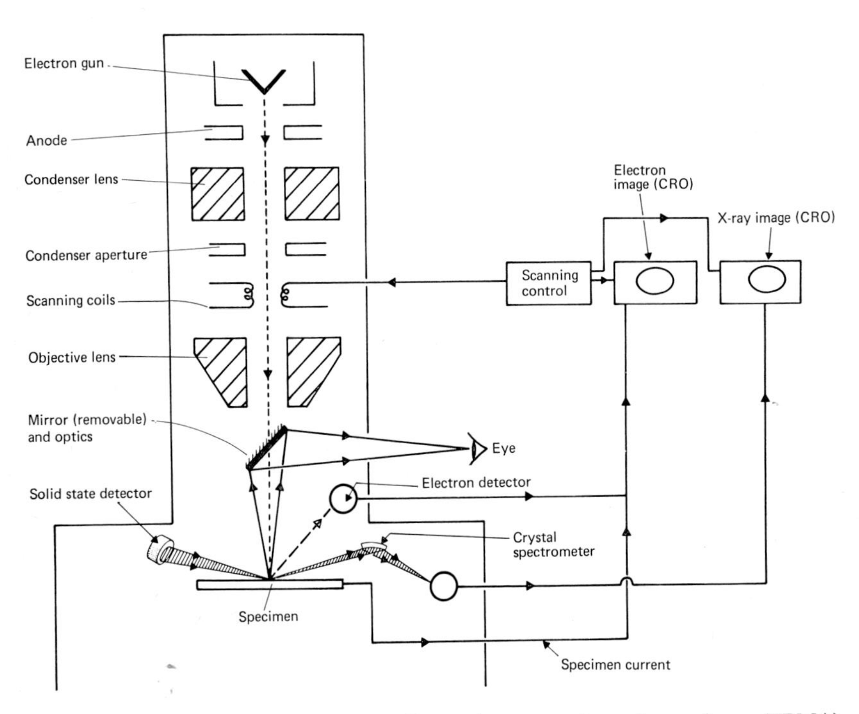

Fig. 3.13. Schematic representation of an electron-probe microanalyser (EPMA). (Courtesy of John Wiley and Sons, Inc., New York.)

The X-rays that emerge from the specimen are detected by one of the three systems already described (§ 3.2). Sometimes, as shown in Fig. 3.13, a wavelength crystal spectrometer is used but more frequently an energy dispersive detector is also employed. The spectrometer and microscope may operate at the same high vacuum or an X-ray transmissive thin plastic window may separate the two. Depending on the type of information that is required from such an instrument it may be operated in a number of different ways.

3.4.1a X-ray scanning image

In addition to the electron scanning image, the X-ray output can also be used to provide an image. The X-ray spectrometer or SSD is set to detect one particular characteristic line (from a single element) and the X-ray output is linked to a CRO. As the primary electron beam is scanned in a raster over the specimen an image can be formed that will be a 'map' representing the distribution of that element across the specimen. The X-ray information generated at each point on the sample modulates the intensity of the moving electron spot on the oscilloscope, so providing a relative intensity distribution. A long persistence phosphor screen allows the image to build up and may be photographed for a permanent record. By simply changing the tuning of the X-ray detector system, distribution maps for different elements may be obtained.

This type of imaging is most useful when high concentrations ($> 5\%$) of elements are present within a specimen. In most biological samples, however, the concentrations of elements other than C, H, O and N are usually less than 1% and such a method of scanning does not generally provide a large enough X-ray signal to be detectable. However, much useful work has been done on material such as teeth and bone where calcium and phosphorus levels are very high. Relatively large aggregates of elements within soft tissues have also been studied in this way.

3.4.1b Linear scan

The electron probe may be made to scan along a line on the specimen surface while the emitted X-rays are being detected, again usually in a crystal spectrometer. A chart recorder linked to the X-ray output may be made to traverse a horizontal line such that the X-ray signal is represented by a vertical displacement of the pen. Thus variations in elemental concentration can be determined in a linear traverse across a selected portion of the specimen.

3.4.1c Static probe analysis

Whereas the X-ray scanning image is chemically specific, it can only be qualitative or at best semi-quantitative. The method involving a linear scan provides some relative quantitation along the line of scan on the specimen, but when accurate quantitative analysis is required, especially when elemental concentrations, and hence X-ray intensities, are low, it is necessary to record the data while the probe is stationary over the chosen specimen area. If the light-optical system or electron signal provides good morphological information from the specimen so that histological detail can be localised, the electron beam may be positioned over a selected area of the specimen and the emitted X-ray signal collected and analysed. A typical counting time in this mode of analysis is 10–100 sec and so low concentrations of elements may be determined. This is the most sensitive mode of operating the EMPA for local analysis.

Often a combination of the three modes of operation is necessary in a particular investigation. A two-dimensional X-ray scanning image indicates the areas of interest on the specimen in terms of their content of elements, a linear scan analysis shows the local distribution of elements, and a point-by-point analysis allows quantitative analysis of small regions, such as subcellular or submicron particles, to be performed. Concentration mapping is also possible by setting the X-ray detector system to receive signals only between certain preset concentration levels, e.g. 5–10%, such that elements present at higher or lower concentrations will not be represented in the oscilloscope X-ray image (Heinrich 1968).

3.4.1d Information from the electron signal

In addition to the X-rays, the various electron signals that are generated from the specimen can also be useful for providing information on elemental distribution. Electrons from the primary beam are scattered by the specimen according to the local atomic number. The larger the atomic number in a particular region of the specimen, the greater will be the degree of scattering, and as far as the backscattered electron signal is concerned the brighter will be that region in the image on the oscilloscope. Thus the backscattered electron image provides a first check on the atomic number distribution within the specimen. The surface topography of the specimen also has an effect on the electron signal. Depressions and asperities on the surface produce areas of shadow and high spots of intensity in the oscilloscope image because the electron detector views the surface obliquely. Both of these types of information may be presented together in the electron image.

If the specimen current is measured, then the display represents regions of high or low electrical conductance and shows up surface irregularities in a different way to that from backscattered electrons. The two signals are often combined to produce more meaningful data about atomic number distribution within specimens.

3.4.1e Light-optical system

Light-optical microscopes with magnifications up to 600× in transmission and reflection are usually incorporated into the EPMA. The installation of a good microscope objective in the region above the specimen is made difficult by the space requirements for the electron optics, but several types of EPMA have a good light-optical system of variable magnification and provision for illumination by transmission as well as polarised light. Because most EPMA instruments are not usually provided with an electron-optical system producing beams narrower than 0.1 μm, and because scanning electron microscopy often lacks adequate contrast in biological preparations, the quality of the light-optical system in the instrument is very important.

3.4.2 Combined scanning electron microscope (SEM) and X-ray detector

The main difference between the electron-probe microanalyser (EPMA) and the combined SEM and X-ray detector is in the design of the electron-optical system. Greater emphasis is placed on the X-ray detection facilities in the EPMA, while the SEM and X-ray system combines these facilities with a much better resolution in the imaging system. Figures 3.14 and 3.15 show a typical arrangement. The image resolution is largely a function of the diameter of the electron probe used and this is often in the range 10–20 nm. Wavelength crystal spectrometers or solid state detectors may be incorporated into the very large specimen region and may be moved close to the specimen to improve collection of X-rays. With the consequent high X-ray detection sensitivity, relatively low electron beam intensities may be employed allowing small probe diameters to be used.

A great advantage of this system is the highly versatile specimen manipula-

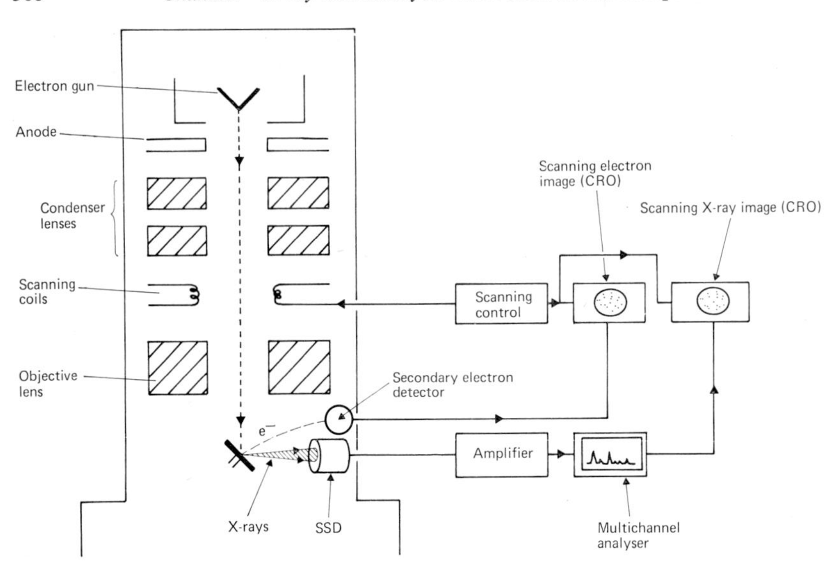

Fig. 3.14. Typical arrangement for a scanning electron microscope (SEM) with a solid state detector attached. Microscopes can also be fitted with crystal spectrometers. (Courtesy of John Wiley and Sons Inc., New York.)

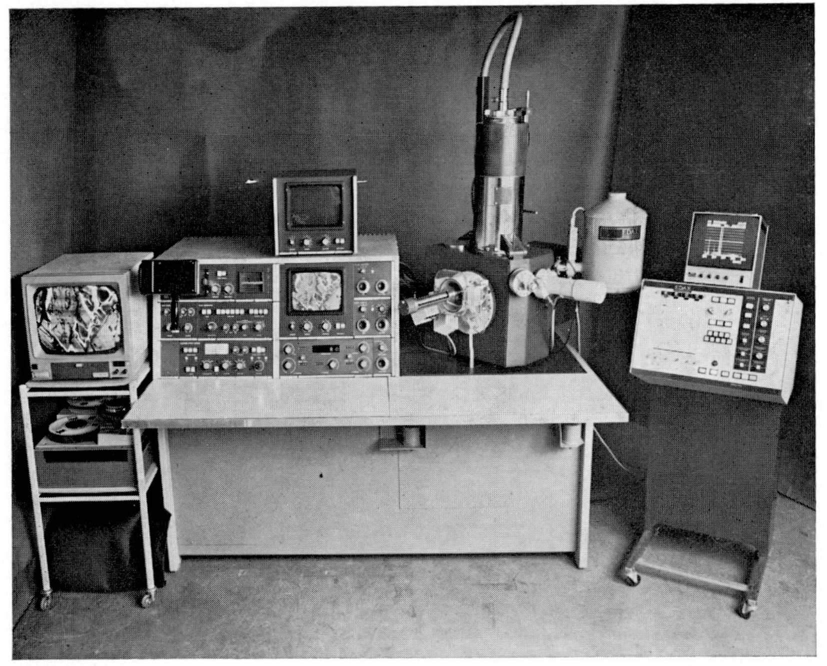

Fig. 3.15. A scanning electron microscope with X-ray detector (SSD) attached. (Courtesy of Pye Unicam Ltd.)

tion facilities that are available for shift, tilt, rotation, etc. as a result of the very large depth of field. The types of analysis (static, one-dimensional, and two-dimensional) and the forms of electron imaging available (backscattered, secondary and specimen current) are similar to those already described for the EPMA (§ 3.4.1).

3.4.3 Combined scanning transmission electron microscope (STEM) and X-ray detector

The *scanning transmission electron microscope* is a relatively new instrument. As well as integral instruments, there are also attachments which can turn conventional transmission microscopes into STEM microscopes. Alternatively, conventional SEMs may be converted to have a STEM capability. Again X-ray detectors may be incorporated into such a system. A simple arrangement is shown schematically in Fig. 3.16. The STEM allows very high brightnesses to be obtained in very small electron probes if careful attention is paid to gun design. In this way, extremely good spatial resolution

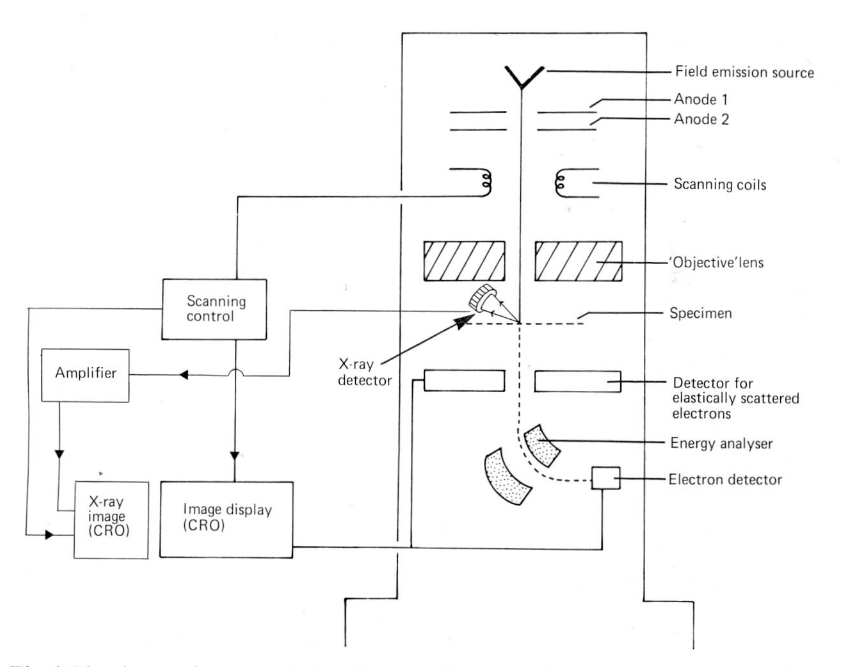

Fig. 3.16. A scanning transmission electron microscope (STEM) with an X-ray detector (SSD) and an energy analyser attached. (Courtesy of John Wiley and Sons, Inc., New York.)

can be obtained during analysis of a thin sample, since a very intense electron beam can be used on a small area of the specimen. Such small electron beam diameters are achieved with a *field emission source* instead of the conventional *hairpin filament* (Agar et al. 1974). Laboratory instruments have been constructed with field emission guns producing probes of 2 nm diameter or less (Crewe et al. 1968; Broers 1967). Suppliers of field emission guns and lanthanum hexaboride cathodes are listed in the Appendix.

3.4.4 Electron microscope microanalyser (EMMA)

A logical extension of the now well-established technique of transmission electron microscopy is to combine high resolution imaging of ultrathin sections with high efficiency X-ray microanalysis. This combination is illustrated in Fig. 3.17 which shows the integral instrument EMMA, which was designed to avoid having to compromise between high resolution

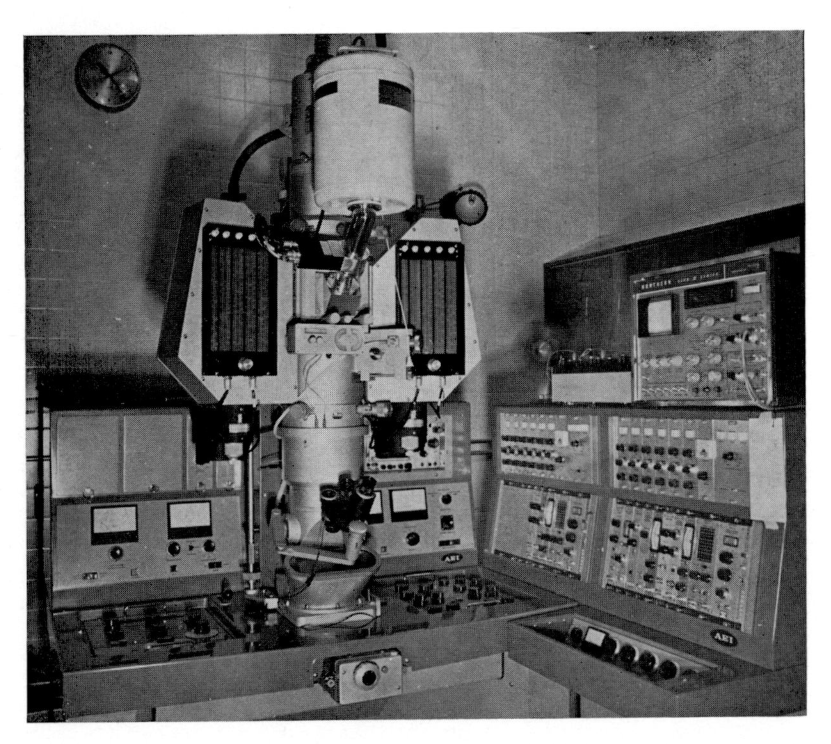

Fig. 3.17. An electron microscope microanalyser (EMMA), fitted with crystal spectrometers and a solid state detector.

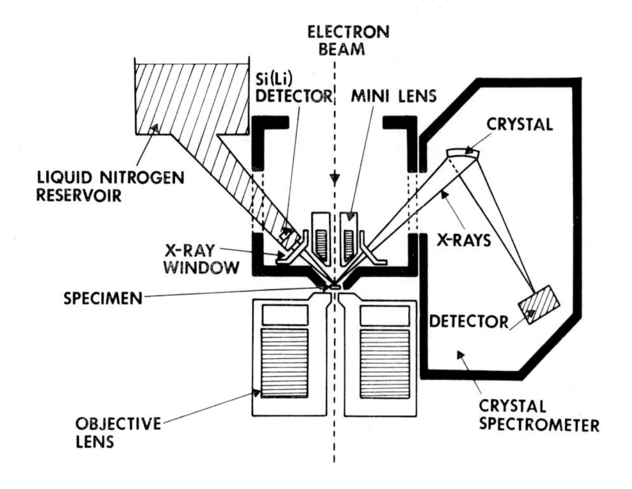

Fig. 3.18. Centre section of EMMA showing the position of the minilens relative to the specimen. (Courtesy of John Wiley and Sons, Inc., New York.)

electron imaging and efficiency of X-ray analysis. The main modification of a conventional electron microscope is in the region immediately above the specimen where an extra condenser lens is introduced (Fig. 3.18). This small conical lens, called a '*minilens*', allows the illumination to be focused to a small probe on the specimen. Probe diameters of 0.1–0.2 μm are commonly employed, although smaller diameters can be obtained with reduced beam currents. The transmission image is visible on the microscope screen even during analysis so that very accurate localisation is possible. X-rays leaving the specimen enter the detector systems (double crystal spectrometers and/or solid state detector (SSD)) via thin windows to allow the microscope to operate at high vacuum. Both the crystal spectrometers and the solid state detector can function simultaneously and in some arrangements the SSD can be made to slide in and out towards the specimen to increase the solid angle of collection of X-rays.

The arrangement for analysis of thin sections is illustrated in Fig. 3.3 which demonstrates the great advantage of this system whereby the ultra-structure of the specimen can be correlated with microanalysis for detailed cytochemical studies. Similar advantages obtain for metallurgical and mineralogical specimens. A chart is positioned on the front of the crystal spectrometers (Fig. 3.19) and allows rapid tuning of the spectrometers with an external gear box drive system for analysis of particular elements.

The electronics associated with the X-ray analysis are visible to the right

Fig. 3.19. Crystal spectrometer on EMMA, fitted with a wavelength indicating chart. The spectrometer is tuned to particular elements by adjusting the drive control. (Courtesy of John Wiley and Sons, Inc., New York.)

of the microscope in Fig. 3.17. They incorporate counting equipment for the spectrometers and multichannel analysis for the solid state detector.

The instrument is operated basically in the same mode for microanalysis as for normal transmission electron microscopy with a range of accelerating voltages from 40–100 kV which is suitable for most types of ultrathin specimen. Although beam deflectors are incorporated in the condenser system the probe is not normally scanned as in the previous three instruments described, the instrument being used essentially as a static probe analyser.

With typical ultrathin sections (< 100 nm thick) the amount of material to be analysed is extremely small (generally $< 10^{-15}$ gm) and such scanning systems would rarely yield useful information. Typical counting times are in the region of 100 seconds per point analysis.

3.4.5 Combined transmission electron microscope (TEM) and X-ray detector

Many existing electron microscopes are capable of being converted into X-ray analysis instruments simply by attaching a suitable detector to the specimen region. This often requires some modification to the microscope column but quite a satisfactory arrangement can often be obtained without too much disturbance of the electron-optical system.

A typical arrangement is illustrated in Figs. 3.20, 3.21 and 3.22. In practice the original design of the microscope objective lens sometimes limits the collection of X-rays to the horizontal direction with the specimen tilted in a goniometer stage. Great care has to be exercised to avoid stray X-ray emission entering the detector from surrounding metal parts of the microscope

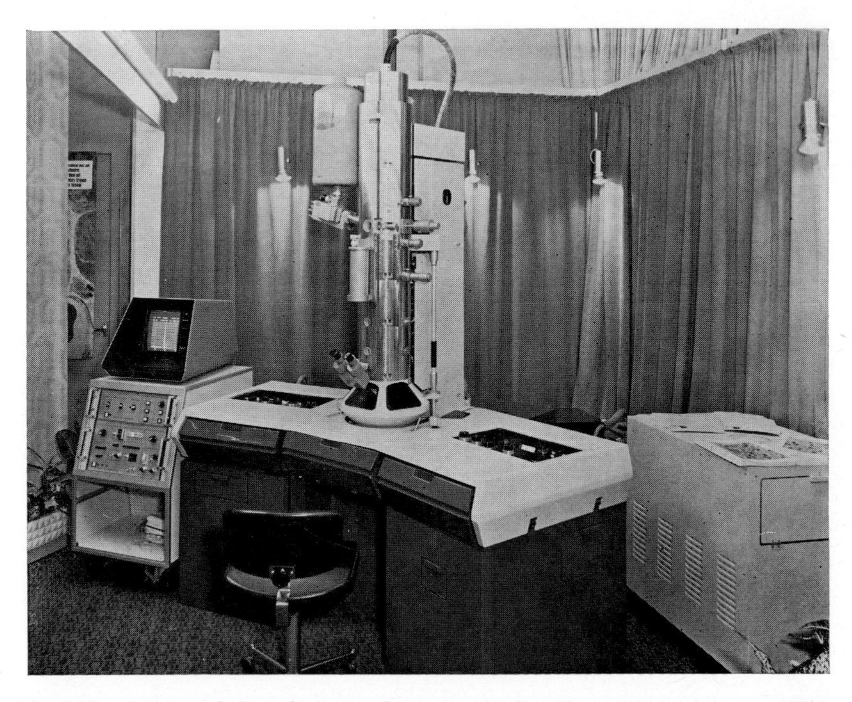

Fig. 3.20. A transmission electron microscope fitted with an X-ray detector (SSD). (Courtesy of Perkin Elmer Ltd.)

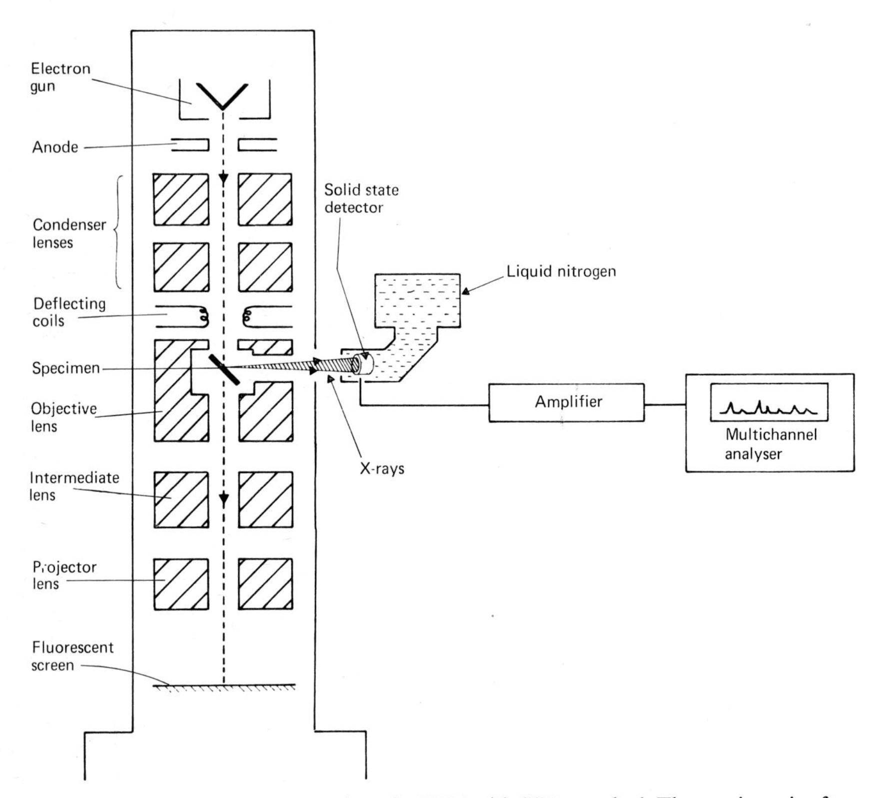

Fig. 3.21. Schematic representation of a TEM with SSD attached. The specimen is often tilted but in some instruments this is not necessary. (Courtesy of John Wiley and Sons, Inc., New York.)

which may be bombarded with electrons. If no minilens is incorporated into the condenser system the smallest available probe is that of the conventional microscope, that is about 1 μm. With a minilens, however, or with careful design of the prefield of the objective lens, probes down to 20 nm can be achieved, although often with low intensity beam currents (§ 5.2.2).

Crystal spectrometers were the first type of detector to be incorporated into transmission microscopes in this way but these have largely given way to solid state detectors. As these detectors become more versatile they can more readily be combined with different electron microscopes and it is expected that there will soon be a great increase in the numbers of transmission electron microscopes with attached X-ray detectors. Again, like EMMA, this instrument finds its greatest use in static probe analysis when X-ray analysis is combined with detailed morphological studies on ultrathin

TABLE 3.4

Relative performance of various electron-optical systems (typical values)

Instrument	*EPMA*	*SEM + X-RAY*	*STEM + X-RAY*	*EMMA*	*TEM + X-RAY*
Specimen type	Thick Thin	Thick Thin	Thick Thin Ultrathin	Ultrathin	Ultrathin
Viewing mode	Light or electron scan	Electron scan (secondary or transmitted)	Transmitted electron scan	Transmission	Transmission
Mag. range (approx.)	50–100,000	20–200,000	20–200,000	200–120,000	200–500,000
kV range	1–30	1–30	1–50	40–100	20–120
Image resolution (nm)	$\sim$ 100	$\sim$ 10	< 1	< 1	< 1
Min. probe dia. for analysis (μm)	> 0.1	0.1[a]	0.1[a]	0.1[a]	$\sim$ 1[b] 0.1[c]
X-ray spatial resolution (μm)	> 1[d]	> 1 (thick) 0.1 (thin)	0.1	0.1	1–2[b] 0.1[c]
X-ray detector	SSD/GFPC/ crystal spect.	SSD and/or crystal spect.	SSD and/or crystal spect.	SSD/GFPC and crystal spect.	SSD or crystal spect.

[a] Typical minimum probe diameter suitable for analysis of low concentrations.
[b] Without minilens.
[c] With prefield objective lens or minilens incorporated.
[d] Depending on kV and specimen thickness.
The above figures represent typical operating values but variations occur depending on the nature of the specimen and on the concentrations being detected.

Fig. 3.22. Inverted transmission electron microscope with SSD attached. (Courtesy of AEI Ltd.)

specimens. Transmission instruments may be operated in both the TEM mode and the STEM mode for analysis when suitable attachments are supplied.

Table 3.4 summarises the relative performance of the various systems described. Typical values are given for the various parameters but these vary from instrument to instrument.

References

Agar, A. W., R. H. Alderson and D. Chescoe (1974), Principles and practice of electron microscope operation, in: Practical methods in electron microscopy, Vol. 2, A. M. Glauert ed. (North Holland, Amsterdam).

Broers, A. N. (1967), Electron gun using long life lanthanum hexaboride cathode, J. appl. Phys. *38*, 1991.

Crewe, A. V., D. N. Eggeberger, J. Wall and L. M. Welter (1968), Electron gun using a field emission source, Rev. Scient. Instrum. *39*, 576.

Hall, T. A. (1971), The microprobe assay of chemical elements in: Physical techniques in biochemical research, Vol. 1A, 2nd Edn., G. Oster, ed. (Academic Press, New York).

Heinrich, K. F. J. (1968), Quantitative electron probe microanalysis, NBS technical note No. 298, Washington D.C., U.S.A.

Russ, J. C. (1972), Resolution and sensitivity of X-ray microanalysis in biological sections by scanning and conventional transmission electron microscopy, Proc. 5th SEM symp. I.T.T. Res. Inst., Chicago, Ill.

Chapter 4

Specimen preparation

The most important condition to be satisfied when preparing specimens for microanalysis in the electron microscope is to maintain the integrity of the specimen. Most methods of preparation involve treating the specimen mechanically, chemically or thermally, and each of these treatments can cause some loss or redistribution of the elements under examination. In addition to displacement or repositioning of elements, contamination can occur even at a level not visible in the electron microscope. Ultimately the aim is to preserve the ultrastructure of the specimen as well as possible while maintaining a faithful elemental composition. Even though the specimen is usually modified during preparation (for example most biological specimens are embedded in a resin and stained with heavy metals) the microscopist must at least be aware of the artefacts that can occur and allow for them.

No attempt will be made here to review the many and various methods for preparing specimens for electron microscopy, but emphasis will be placed on the modifications that are important for X-ray analysis. Space limitations prevent the inclusion of all techniques which have been used successfully for analytical work. Full details of specimen preparation techniques are given in other books in this series (Goodhew 1972; Glauert 1974; Reid 1974; Lewis and Knight 1977) and in the papers suggested for further reading given at the end of this chapter. Again it stressed that the scope of this book does not include the preparation and analysis of bulk materials (i.e. 'infinitely' thick specimens) but concentrates on the techniques used for the preparation of thin specimens that may be examined in the SEM, STEM, TEM and EMMA.

4.1 Specimen supports

4.1.1 Thin specimens (TEM)

Specimens are mounted on rigid supports (grids or rings) which may be covered with a very thin (10–20 nm thick) supporting film, or may be used bare. The choice of material for the support depends on the type of specimen being analysed but general rules govern which type of support to use.

A number of types of electron microscope grid for supporting specimens are available, varying in mesh size, format and composition. Grids are made from copper, nickel, gold, molybdenum, stainless steel, aluminium, nylon (carbon coated), beryllium, titanium, chromium and platinum (see list of suppliers in Appendix). Because of the electron scatter which occurs during electron bombardment of the specimen, (see § 5.1.4), spurious X-ray lines emitted by the grid material may enter the X-ray detector. Thus if copper is being analysed in a specimen, a copper grid cannot be used, and a nickel or aluminium grid is preferred. The operator must be careful to choose the correct grid material that will not interfere with the analysis. A quick check in a book of X-ray tables (e.g. Bearden 1964) will show if spurious lines are likely to occur.

As an example, assume the operator has a specimen in which it is required to detect and measure Na. The K_α line of sodium (at 1.041 keV) is the most readily measured but if the specimen is mounted on a copper grid or support then a spurious line of CuL_β at 1.023 keV may be produced. If the level of Na is very low then the possibility of confusion exists between the two lines, even though the copper line is of very low intensity. It would be better to mount this specimen on an Al grid.

Nylon grids have been found to be very useful for analysis since virtually no X-ray characteristic emission is produced. Some nylon grids coated with carbon do produce a small TiK_α characteristic line due to a contamination level of about 1 % of titanium. Another possible disadvantage of these grids is that they are in the form of a woven mesh rather than being photo-etched, and hence are less stable and not completely flat. They do, however, have the advantage of producing a very low X-ray background signal (see § 6.2.1a). Perforated beryllium foil, in the form of discs 3 mm in diameter, has been used by the author with some success for specimen supports, as have carbon rings.

The largest possible grid mesh should be chosen since the electron scattering effect is less likely to be troublesome when the specimen area is a greater distance from the grid bars (Chandler and Morton 1976). Numbered 'finder' grids (see Appendix) are extremely useful for mapping regions of the specimen for subsequent analysis (Henderson et al. 1973). Aluminium grids are often difficult to handle since they have a tendency to buckle.

Supporting films for use in the TEM are generally of carbon, nitrocellulose or polyvinyl formal (Formvar). They must be thick enough to support the specimen (biological or mineral powder usually) and to conduct heat effectively, yet thin enough to allow adequate electron penetration without loss of resolution. Typical thicknesses range from 5 nm to 30 nm. Some supporting films contain significant traces of silicon, chlorine and sulphur which must be measured independently during the analysis of these elements. Carbon supports, evaporated *in vacuo* from spectrochemically pure graphite rods, are extremely 'clean', and produce no spurious X-ray lines, but are relatively less strong than some plastic films. Methods for the preparation of specimen supporting films are described by Bradley (1965) and Goodhew (1972).

4.1.2 Thick specimens (SEM)

For analysis of thicker samples in the SEM (or EPMA) a carbon or beryllium substrate is often used to mount the specimen. Both of these elements produce no troublesome characteristic X-ray lines and very low X-ray background. Specimens are sometimes coated with an evaporated film of carbon or aluminium to reduce electrostatic charging and heat effects in the specimen, although this is less frequently necessary in metallurgical and mineralogical samples than in non-conducting specimens, such as biological tissue. Specimens may be fixed to the substrate with a colloidal graphite or silver paste, or with an adhesive such as Bostik which must then be coated with carbon to make it conducting (see Appendix).

The specimen must be mounted on a very flat, and again highly polished substrate. Glass slides are often used when correlated histological observations of biological specimens are required, although it is preferable to use slides of fused quartz of high purity since glass may contain elements (e.g. Na, Mg, Ca) which may interfere with the analysis. For the analysis of silicon, the specimen is mounted on a neutral substrate such as a plastic or on a block of carbon.

Thin films of carbon, nylon, or collodion on electron microscope grids or supporting rings are preferable as supports for sections where the electron beam may penetrate through to the substrate. Such supports produce less background from white radiation than do solid supports. Beryllium or pure carbon substrates have also been used to reduce the background radiation. For thicker specimens, where electrons do not reach the substrate, metal stubs can be used.

Hall et al. (1972) and Echlin and Moreton (1974) describe the following method of producing nylon films 100–200 nm thick to act as supports for thinner specimens where low background is essential:

(i) Dissolve 50 gm of nylon chips in 500 ml of isobutyl alcohol by gently heating in a liquid paraffin bath to 90 °C. Keep as a stock solution.
(ii) Place suitably sized aluminium collars on a metal plate in a shallow tray (5 cm deep × 60 cm × 30 cm) filled with cold clean water.
(iii) Ensure there are no air bubbles in the collars and between the aluminium plate and the collars.
(iv) Cool the dissolved nylon mixture to 60 °C.
(v) Pipette a stream of nylon mixture, equivalent to 10 drops, onto the water surface and allow to solidify as a film.
(vi) Remove and discard the film with a glass rod.
(vii) Pipette a stream of nylon mixture, equivalent to 4 drops onto the water surface and allow to spread and solidify.
(viii) Manoeuvre the aluminium collar under the film so that it is immediately beneath a purple or gold coloured region.
(ix) Drop the water level (by tap or pipette) so that the film lowers onto the collars on the plate.
(x) Store the coated collars in a dessicator to dry.
(xi) Evaporate a thin layer of aluminium or carbon onto the nylon films to provide a conducting layer.

Sections for analysis are now mounted onto the collars and if necessary coated with carbon or aluminium as described below. Echlin and Moreton (1974) also describe methods of mounting frozen sections or cell suspensions.

4.1.3 Specimen coating

In order to reduce damage to thick specimens (SEM) during electron bombardment they may be coated with films of conducting materials such as aluminium, carbon or copper. Obviously a material must be chosen that will not interfere with the analysis. Films of between 5 and 50 nm are commonly used and are deposited by evaporation *in vacuo* onto the mounted specimen. These thin films allow the heat and electrostatic charge generated in the specimen to be conducted safely away. The film thickness must be carefully chosen to avoid absorbing low energy X-rays. Methods of monitoring film thickness are described by Goodhew (1972). Such films are necessary, not only for reduction of damage, but also to allow good quality scanning electron images to be formed (Thornton 1968). With thick specimens the whole of the sample needs to be coated by rotating and tilting it during the evaporation procedure. These coating techniques are described by Echlin (1974).

Metallurgical and mineral specimens do not usually require a conducting film when a large area is being examined. However, a dispersion of small particles may require stabilisation in the electron beam to avoid elemental loss and conducting coats may be necessary. For most biological samples a coating film is almost always necessary.

Thin specimens (TEM) are coated with much thinner layers ($\sim$ 5 nm) than thicker specimens to avoid decreasing the resolution of the image. Methods of depositing thin films *in vacuo* are described by Holland (1956) and Goodhew (1972).

Non-conducting thick samples for examination in the SEM may also be painted at the edge with a small streak of Aquadag colloidal silver or colloidal graphite to provide better electrical contact with the supporting base.

4.2 *Metallurgical specimens*

4.2.1 *Preparation of specimens for analysis*

In general there is very little difference in the preparation methods used for X-ray analysis of metals compared with those for conventional electron microscopy. The normal hazards of specimen changes, for example by heating, are just as much a concern to the electron microscopist as to the analyst. Many methods of thinning impart energy to the specimen and, whereas conducting specimens may be little affected, the deformation and structural changes in non-conductors may be very serious. The analyst must also be aware of the possibility that material may be added to, or subtracted from, the original specimen. A full description of the techniques for preparing metallurgical specimens is given by Goodhew (1972), who describes methods for the preparation of bulk specimens prior to final thinning by electro-polishing, chemical polishing, ion-beam thinning, flame polishing, ultra-microtomy and mechanical methods. Techniques are also described for the preparation of materials not in sheet form, such as extraction replicas, powders, wires and fibres.

Some of the special problems to be aware of in the preparation of specimens for analysis will now be discussed. The precautions are similar for both thick (SEM) and thin (TEM) specimens and are considered together.

4.2.2 Chemical changes during preparation

The removal of second phase material may occur during polishing with the subsequent creation of holes. If the specimen is being mechanically polished then there is the risk of these spaces being filled with polishing compound, which may contain Al_2O_3, MgO or SiC. Thus the ultrasonic cleaning or washing of specimens in a suitable solvent is necessary before analysis. Inadequate washing also often leaves behind residue which may or may not be visible in the electron microscope image but which could interfere with the analysis. The preparation of extraction replicas may occasionally involve the deposition of a layer of heavy metal on the specimen for shadowing purposes. Such layers may be of platinum-rhodium alloy, gold, gold-palladium alloy or tungsten oxide. Whatever metal film is used it will be necessary to ensure that the X-ray emission from the shadowing layer does not interfere with the analysis of the specimen.

4.2.3 Specimen thickness (TEM specimens)

Specimen thickness variations can cause errors in quantitative analysis and these are discussed more thoroughly in Chapter 6. Briefly, in a metal foil of varying thickness it may be difficult to determine relative elemental content across the specimen. It is then necessary to use the mass thickness measurement technique described by Hall (1971), or the ratio method described later, in Chapter 6. Wherever possible, analyses should be performed on a part of the specimen having approximately uniform thickness, as indicated by the transmitted electron beam current for a given specimen composition, or by the white radiation. The measurement of metal film thickness by X-ray microanalysis has been described by Hutchins (1966) and Chandler (1976). Lorimer et al. (1975) described the measurement of specimen thickness by depositing contamination spots on both sides of the specimen. The specimen is subsequently tilted and the separation between the spots is measured. The method has especial value when thicknesses are in the range 100–1000 nm, and makes use of the parallax phenomenon.

The actual specimen thickness is often of more importance in the study of metallurgical specimens than of biological ones. The consequences of X-ray absorption and fluorescence are more serious for thicker specimens, and it is also preferable to have as thin a specimen as possible for ultrastructural studies. Broadly speaking, the criteria for selecting the thickness of metal specimens for high-resolution electron microscopy also apply in X-ray analysis. The specimens must be thick enough to provide a suitable X-ray signal for detection, yet thin enough to allow adequate image resolution, to avoid absorption and fluorescence, and to give good spatial resolution of analysis (see § 3.1.1). The critical thickness for true 'thin' specimen microanalysis can be established using the method described by Cliff and Lorimer (1972). The ratio of the intensities of the characteristic line of the element to be analysed and the white radiation measured in a band of the spectrum containing no spectral lines, is plotted as a function of specimen thickness. The intensity of the characteristic line is taken as a measure of the mass of the element (§ 6.2.1) concerned and the white radiation as a measure of the specimen mass thickness (§ 2.4.4, § 6.2.1). As the specimen thickness reaches a point at which absorption of the characteristic line occurs, the line deviates from linearity. As a rule (Cliff and Lorimer 1972) the limit of a 'thin' specimen occurs when the material is not transparent to 100 kV electrons, but this depends on the energy of the elemental characteristic line being measured and the matrix in which the element is incorporated.

Jacobs and Baborovska (1972) have shown how the thickness of a metal foil determines the ratio of two characteristic line intensities from a thin specimen of a homogeneous alloy. The higher the characteristic line energy, the greater may be the foil thickness before absorption occurs.

4.2.4 Specimen thickness (SEM specimens)

Metal specimens that are thick enough to entirely attenuate the electron beam (a few micrometres at 50 kV accelerating voltage) are here termed 'thick'. The analysis of this type of specimen falls into the range of general electron-probe microanalysis which is not considered in detail here. A wealth of literature provides descriptions of this type of analysis (see Chapter 1) which sometimes involves quite complex correction procedures for quantitative work and generally employs the EPMA.

A particularly important consideration in the preparation of thick specimens for SEM is the smoothness of the surface to be analysed. The difficulties involved in the analysis of specimens with irregular surfaces are similar to those for hard biological material (§ 4.4.1d) and are illustrated in Fig. 4.1. Metal specimens in the form of discs that have been turned on a lathe, for example, will not have the requisite smoothness and must be mechanically polished with a suitable abrasive (of a grade less than 1 μm) and then thoroughly washed in alcohol, or finally electropolished (Goodhew 1972). Electropolishing, however, may cause the preferential surface removal of phases in certain alloys. A review of metal surface analyses with emphasis on specimen preparation difficulties is given by Hutchins (1969).

4.3 Mineralogical specimens

As with metallurgical specimens, the preparation of mineral samples for X-ray analysis in the electron microscope closely follows conventional methods for normal imaging. Mineralogical specimens may be prepared from bulk and thinned down, or may be in the form of a powder, dispersed onto a support film, for studies of individual particles.

4.3.1 Thinning from bulk specimens

Ion-beam etching of specimens is a useful method of thinning minerals from the bulk, as described in detail by Goodhew (1972). A beam of ions is directed onto the specimen until part of it becomes thin enough for electron transmission. Possible artefacts that may interfere with accurate analysis of this type of specimen are thickness variations, which may affect quantitation; absorption of X-rays in the specimen; crystal lattice orientation (Duncumb 1962), which may affect X-ray intensity; and ion migration during the analysis (see § 5.1.1).

4.3.2 Powder specimens

The most common method of examining minerals in the electron microscope is in powder form. Size ranges from 5–5000 nm are usually encountered and particles as thin as 10–20 nm have been successfully analysed. In order to examine single particles they must be dispersed from the powder onto a supporting film mounted on an electron microscope grid. Adequate dispersion of particles is important for the analysis. If particles in a powder dispersion are lying too closely together (e.g. < 1 μm apart) on the support film, then during the analysis of one particle electrons may be scattered from it to interact with the neighbouring particle, so producing X-rays from both particles together. To avoid this, the dispersion should provide a distance of at least 1 μm between particles.

A simple way of dispersing particles and preventing aggregates on the support film is to suspend a small amount of the powder in a test-tube in about 10–100 times its volume of methanol (usually about 2 ml). The solvent should be Analar grade. The powder is dispersed by ultrasonic vibration, and then a 5 μl droplet of the dispersion is taken up in a micropipette (see Appendix) or carefully in a thin pasteur pipette, or even on the end of a needle, and placed on the support film on the grid. The droplet is then dried in air or held under a 40 W table lamp to dry more quickly. Instead of methanol, ethanol or even deionised or distilled water may be used for dispersion (not tap water!). Care has to be taken with water, however, since many minerals will be soluble or may lose their crystal structure in water.

Some workers prefer to use a spraying technique to form an aggregate-free dispersion of powder on the microscope grid. In this method, described by Goodhew (1972), a clean glass slide is first sprayed with a suspension of the particles and then coated with a film of carbon. The carbon film with adhering particles is then removed from the slide by floating on distilled water, and mounted on a microscope grid.

Alternatively a dispersion of the powder in methanol may be sprayed directly onto a number of electron microscope grids already coated with support films and resting on filter paper. The grids will then be ready for immediate examination (after perhaps a further coating of carbon by evaporation).

Some powdered oxides aggregate even in alcohol and hydrochloric acid has sometimes been used as a deflocculent. A review of techniques for preparing powder dispersions is given by Comer (1971).

4.4 *Biological specimens*

X-ray microanalysis has been performed on both hard and soft tissue, in thick and thin forms, as well as on isolated cells. In an interpretation of the analytical results the investigator must have full confidence in the integrity of the specimen; that is that the elements that have been analysed are distributed within the specimen in the same way as in the original living tissue, or at least that any differences should be entirely understood. Different preparatory techniques will affect elements in the tissue in different ways depending on whether they be free or bound. Soluble electrolytes for example (Na, K, Cl) are unlikely to remain in their *in vivo* situations when tissues are immersed in organic solvents or aqueous solutions. The binding forces linking elements to organic structures, such as proteins, are very complex and little understood. Elements may exist also in tissue in both the bound form and as free ions and will be affected differently by different procedures. Correlative techniques are necessary to establish the nature of these binding forces and the effect of various treatments on them. For example, atomic absorption spectometry may be used to determine the degree to which elements are free or bound in tissues subjected to conventional preparative procedures involving solvents.

X-ray microanalysis has been used in studies of a very wide range of biological problems and a great diversity of specimen preparation techniques have also been used. Consequently it is not possible to list all these methods here. For many applications they follow conventional fixation and embedding procedures, and routine histochemical methods, while for other applications it is essential to use freezing techniques.

These methods are described in detail in other books in this series (Glauert 1974; Lewis and Knight 1977) and so the special preparative methods for X-ray analysis will only be considered here in terms of actual applications. Attention will be drawn to some of the most important problems and areas of difficulty. The reader is encouraged to refer to the literature for detailed descriptions of the methods used in different types of application.

Specimens for analysis may be classified as thick or thin when considering preparation methods, and the procedures themselves may be categorised as those involving solvents, those employing freezing methods without solvents, and those for cell suspensions, incorporating air-drying. In this context, all specimens greater than 200 nm are termed relatively ‘thick’. Specimens less than 200 nm may be examined by transmission and are termed ‘ultrathin’.

4.4.1 Thick specimens (primarily for SEM)

The need for preservation of the finest details of ultrastructure is not so great in the examination of thick specimens by SEM as for transmission electron microscopy since the resolution is limited by the effects of electron penetration and diffusion as described previously (§ 3.1.1). However, care must still be paid to factors affecting elemental distribution, and in an effort to overcome these problems a number of special preparation techniques have been developed.

4.4.1a Routine fixation and embedding

Standard histological methods of fixation, embedding and sectioning are suitable for some types of analysis of thick specimens provided it can be shown that the final specimen is sufficiently similar to the original tissue. Pearse (1968) discusses in detail the factors affecting the preservation of tissue components during preparation for histology. These histological methods involve the use of fixatives such as alcohols, aldehydes, or acetone; embedding media such as paraffin, styrene or methacrylate; clearing solutions such as xylene, toluene, benzene, chloroform or cedarwood oil; and possible flotation of sections on water and staining for histological correlation or histochemical tests. Any of these procedures may contaminate the specimen or cause elemental loss or redistribution. In several experiments it has been reported that the losses have been very great, while in others on different tissues and with analyses of different elements, the changes in the distribution of elements have been acceptable.

Although there are many examples in the literature of analyses performed on conventionally fixed, dehydrated and embedded material, a good deal of care must be exercised in the interpretation of results obtained from such analyses. The diffusible ions likely to be found in solution in the living system are extremely difficult to stabilize in the prepared specimen. A satisfactory analysis of a fixed specimen demonstrating a distribution of elements does not necessarily indicate that the integrity of the specimen has been preserved. For example, Andersen (1967) demonstrated that preparation of red blood cells (of the *Amphiuma*) by chemical fixation in glutaraldehyde followed by OsO_4 caused a large loss of sodium and phosphorus when compared with samples prepared by simple drying methods. The author (unpublished data) has found a similar effect with human sperm cells which lost Na, Cl, S, P and K during treatment with water and with glutaraldehyde. Similarly, Hall and Höhling (1969) analysed the calcium and phosphorus content of rat dentine after fixation in ethanol and embedding in methacrylate and demonstrated a loss of about 75% of the phosphorus content when compared to frozen sections (4 μm thick).

Robison et al. (1971) made a careful analysis of calcium, iodine and phosphorus distribution in human thyroid glands by EPMA. The tissue was fixed in calcium-free and phosphorus-free formalin (3% formaldehyde) and then treated in two different ways. One part was frozen, freeze-dried, embedded in paraffin and cut into slices, 12 μm thick, and then cleared, while the other part was dehydrated with ethyl alcohol, embedded in paraffin wax, sectioned at 12 μm and cleared and the sections mounted on quartz supports. The two methods showed essentially the same content and distribution of iodine, phosphorus and calcium throughout the gland. A third method involved direct freezing with no fixation, followed by sectioning at 12 μm on a cryostat microtome before freeze-drying. Calcium concentrations in these frozen sections and in sections of formalin-fixed material were found to be the same within the accuracy of the analysis. The authors concluded that the calcium was bound to the tissue and not distributed in a soluble form. The distribution of calcium as recorded in the EPMA was very similar in both types of section of many human and animal thyroids, also indicating that calcium was not redistributed during fixation within the limits set by the resolution of the image. However, in sections of 12 μm thickness, this resolution, as limited by electron diffusion, could well have been several micrometres.

If a proportion of an element is bound in the tissue, and part exists as free ions, then the solvents used may have the effect of separating out the free from the bound, depending on their solubility in the solutions used.

Several analyses of elements which do not occur naturally have been performed and for these conventional methods of specimen preparation may be entirely suitable.

The distribution of hard inorganic particles in tissue is usually little affected by chemical treatments and it is then possible to correlate morphology, perhaps studied in previous histological examinations, with local elemental analysis. In certain hard calcified tissues elemental losses may be less severe during processing than in soft tissues (Höhling and Nicholson 1975).

The thickness of the section influences the analysis since heat conduction during electron bombardment is better from a thin section (< 2 μm) having good thermal contact with its substrate. Also the effects of electron diffusion (§ 3.1.1) and electron interaction (§ 5.1) are important. The section should be of uniform thickness since fluctuation in thickness can cause changes in the X-ray intensity which may lead to confusion regarding elemental concentrations. Hall (1968, 1971) and Marshall and Hall (1968) have described the effect of thickness variation on the X-ray output from a thick specimen and have proposed a quantitative method of compensating for it (see Chapter 6).

Studies on sections of embedded material are possible when the embedment used is stable in the electron probe, but one must be aware of the limitations in specimen integrity and in quantitation. Suitable embedding materials include epoxy resins and methacrylates. Cryostat sections of unembedded material provide more reliable data, however, (see § 4.4.1b and c). Sections which have been cleared of embedding medium may behave like cryostat sections as far as quantitation is concerned (Hall 1971), but the clearing process introduces further risks of elemental loss or redistribution.

Methods involving histochemical procedures have been used effectively to stain thick sections for analysis. In this way, the SEM and EPMA have been used to demonstrate the specificity of histochemical staining techniques. Since the analysis depends only on the detection of single elements, the technique does not depend on the colour or density changes in the specimens required for light microscopy. Thus a much wider range of stains is available for analysis of tissue components. Many of these are listed by Pearse (1968). There is no rule of thumb governing the preparation of thick specimens for analysis by histological techniques. A careful check on the elemental concentration of the solutions employed, both before and after treatment, will show whether loss or contamination has occurred, but wherever possible the use of solvents should be avoided.

4.4.1b Freezing methods

A commonly employed technique is to rapidly freeze the tissue block and then freeze-dry it in vacuum until all the water is removed. The tissue is then infiltrated with embedding material ready for sectioning. This method reduces the risk of sample contamination, loss of soluble substances, displacement of cell constituents and chemical alterations of reactive groups during fixation and dehydration (Lauchli 1967). If the tissue is plunged into the low temperature bath immediately upon removal from the living organism, all chemical processes are abruptly halted. Rapid freezing (quenching) of the whole tissue is essential to reduce ice-crystal formation, and the smallest possible size of tissue fragment (1–2 mm^3) is chosen to allow the whole sample to cool quickly. Tissues are quenched in liquid nitrogen or in isopentane, Freon, propane or ethylene glycol, cooled with liquid nitrogen. These provide better thermal contact with the tissue than liquid nitrogen alone. In liquid nitrogen a layer of gas bubbles may form around the tissue surface causing a reduction in the cooling rate.

After cooling, the tissue is freeze-dried in a vacuum apparatus where it is maintained between $-30\,°C$ and $-70\,°C$ and at a vacuum between 10^{-3} and 10^{-5} Torr. Drying can take anything from up to a day or more depending on the size of the tissue block. Various types of commercial freeze-drying apparatus are available (Pearse 1968).

After freeze-drying the tissue is slowly thawed and vacuum embedded by allowing it to sink into a bath of low viscosity embedding material as the temperature is raised. When the vacuum is released the positive pressure aids impregnation of the tissue with the medium. Commonly used embedding media are methacrylates and epoxy resins which do not have to be removed (unlike paraffin wax), since they are relatively stable under electron bombardment. An account of various embedding media is given by Drury and Wallington (1967).

A method of freeze-substitution described by Lauchli et al. (1970) for botanical specimens can also be applied to some animal tissues. The frozen tissue is dehydrated by immersion in anhydrous ether for a period of several days, followed by infiltration with a low viscosity epoxy resin embedding medium (Spurr 1969). Sections 1–2 μm thick are floated on hexylene glycol instead of water to avoid leaching of electrolytes. However, the possibility of contamination still exists at this step, as previously mentioned, through the introduction of elements such as S, Cl, Si or Ca from the embedding medium.

When attempting the localisation of readily diffusable electrolytes such as Na, Cl and K, fixation and staining of the tissue block by exposure to osmium tetroxide vapour has been performed as an intermediary step between freeze-drying and embedding (Ingram et al. 1972). Analysis of 3 μm sections in the EPMA demonstrated some success in retaining these elements and localising them intracellularly and extracellularly.

Pearse (1968) reviews freezing methods in detail.

4.4.1c Cryomicrotomy

In order to minimise the movement of elements within the tissue, specimens can also be prepared by cryomicrotomy. Several approaches to this technique are discussed in detail by various authors in the book edited by Roth and Stumpf (1969) and in papers by Echlin and Moreton (1974), Moreton et al. (1974) and Saubermann and Echlin (1975) and only a general account is given here.

The block of tissue is first quickly frozen as described above, and is then sectioned directly in a cryostat without any embedding. A small layer of minced tissue, such as liver, may be spread on the top of the specimen holder underneath the specimen to act as a cement when the holder is plunged into the quenching medium. The frozen specimens can be stored in liquid nitrogen until required. Sectioning is usually performed between $-30\,°C$ and $-85\,°C$, and then the frozen sections are transferred to a suitable freeze-drying apparatus for about 24 hr. Sections down to 1 μm may be cut by this technique using a cryostat but 5 μm sections are more conventionally prepared. If there is no freeze-drying apparatus within the cryostat, the frozen sections are transferred to a suitable freeze-drier via small precooled plastic containers. When dried the sections are brought up to room temperature slowly for analysis and stored in a desiccator or are analysed frozen in a cold stage (Echlin and Moreton 1974).

The structure of the tissue in these sections is often distorted as a result of ice-crystal formation and of uneven compression and folding during sectioning. The morphological information is nowhere near as precise as with conventional methods of fixation, embedding and staining, but is often good enough for the relatively low resolution analysis of thick specimens. Many of the techniques for cryomicrotomy are reviewed by Pearse (1968).

In a similar method, Gehring et al. (1971) prepared sections of frog skin for analysis by freezing at $-150\,°C$ in isopentane and cryo-sectioning at $-50\,°C$ with a steel knife at a thickness of 2.5 μm. A collodion-covered silver grid was pressed onto the section on the knife and the mounted specimen transferred to a freeze-drier. After drying another collodion-covered grid was placed on top of the section so that the section was sandwiched between the two support films and grids. The section was thus effectively protected from the atmosphere during transfer to an SEM.

The removal of water by freeze-drying causes no loss of elements from the tissue, but by the nature of the drying process, where an ice-water vapour front moves through the tissue, there exists the possibility of some redistribution of soluble substances.

In instruments fitted with cold stages, thick sections have been analysed in the frozen hydrated state without freeze-drying (§ 7.3.4). This is a relatively new approach but may yield important information from specimens kept in a state as near as possible to the original living conditions.

4.4.1d Polishing

If a hard tissue such as bone is to be analysed without sectioning, the surface must be very flat and highly polished to the sub-micrometre level. Absorption of X-rays by asperities (Fig. 4.1) can cause large errors in local analysis, especially if the X-ray collecting angle is shallow.

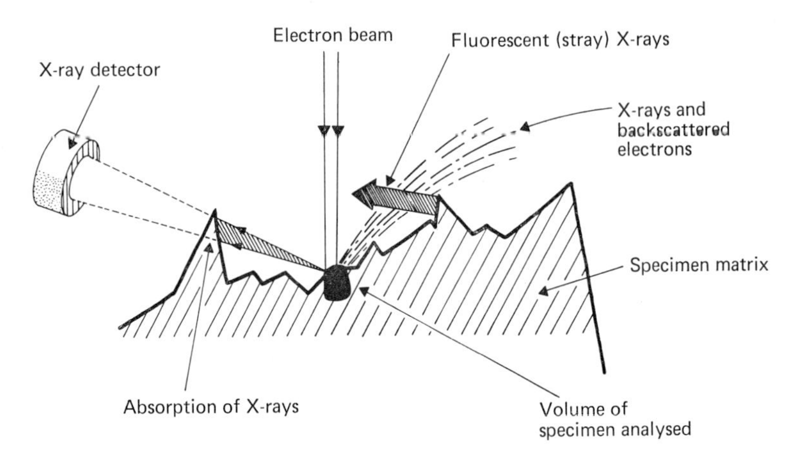

Fig. 4.1. The effect of surface roughness on X-ray collection. (Courtesy of John Wiley and Sons, Inc., New York.)

In the study of mineralised tissue in the EPMA or SEM the specimens are often polished after routine histological embedding and sectioning. However, great care has to be exercised in the process to avoid contaminating the specimen surface with abrasive material. After polishing there may still be surface irregularities due to differences in resistance to abrasion on the specimen surface. This is most serious at junctions between hard and soft tissue (Boyde and Switsur 1963). A diamond saw of diamond size 1 μm is sometimes used for sections of thickness 100 μm (Andersen 1967). Fractured surfaces of mineralised tissues may be analysed but suffer the same difficulties of surface roughness. A review of specimen preparation techniques for hard tissues is given by Höhling and Nicholson (1975) who recommend the use of cryostat sections.

4.4.1e Air-drying

When performing analyses on cell suspensions, such as blood or sperm cells, a relatively simple air-drying procedure can be adopted. A drop of the cell suspension is placed on the supporting stub or smeared onto a slide or coated grid and allowed to dry under a warm stream of air. Alternatively the drop can be left for a minute or so and drawn off with the corner of a filter paper to leave a smear behind. The cells are coated with a conducting film as described earlier (§ 4.1.3).

Some care has to be taken to obtain 'pure' cells by this method. For example, human semen contains much debris in the seminal fluid besides the sperm cells themselves, some of which can dry down onto the cells and contaminates them. Washing the cells before drying down is not satisfactory because this may cause a flux of electrolytes across the plasma membrane (Maynard et al. 1975). There is also a risk of an ionic flux occurring as the liquid-air interface sweeps through the specimen during drying.

It has been shown that no flux of ions occurs during air-drying of sperm cells (Chandler and Battersby 1976a). Subcellular fractions, such as cell nuclei can also be prepared in this way for analysis in the SEM and TEM. Unfortunately a collapse of cell structure occurs in the air-drying process and serious morphological damage is often seen. Polliack et al. (1973) compared the air-drying procedure with methods of critical point-drying. For blood cells the latter method provides a superior preservation of morphology, but ionic fluxes occur as described above. Freezing or freeze-drying of cell suspensions may help to overcome this problem (Chandler and Battersby 1976a). The difference in analysis of specimens prepared by various methods of freezing and freeze-drying is discussed by Gullasch and Kaufmann (1974).

4.4.2 Thin specimens (primarily for TEM)

In the preparation of ultrathin sections (< 200 nm) to be analysed by transmission electron microscopy in EMMA, TEM or SEM instruments, the same principles apply regarding elemental loss and redistribution as for thick sections, but the methods involved in overcoming these difficulties vary somewhat.

Since the aim of performing analysis on ultrathin sections is that of correlating the ultrastructure of the specimen with the analysis, a double problem arises. Methods must be chosen that will not affect the distribution of elements and will also preserve the fine structure of the specimen. With ultrathin sections the versatility of the analytical techniques is enlarged. Conventional methods of preparation for TEM can often be employed, but resort to freezing techniques is sometimes necessary, and methods of histochemistry, enzyme histochemistry and immunohistochemistry allow elemental localisation and molecular identification on a very fine scale. There is no general rule for choosing a preparative method for analysis. Conventional techniques may be attempted, but if unsatisfactory, further investigations with methods avoiding solvents must be made.

4.4.2a Conventional methods of fixation and embedding of tissues

(i) *Fixation* A full account of procedures for the preparation of tissues by fixation is given by Glauert (1974).

The fresh tissue is diced into cubes ($\sim$ 1 mm^3) and placed in buffered fixative, such as glutaraldehyde, formaldehyde, osmium tetroxide, or potassium permanganate. The aims of the fixation are to preserve the cell structure as close to the living state as possible and to protect it against disruption during embedding, sectioning and exposure to the electron beam. Different fixatives are chosen for optimum fixation of different cell components and a mixture or sequence of fixatives is often used to achieve good all round preservation. Glauert (1974) discussed the factors affecting the quality of fixation such as pH, concentration, specimen size, temperature and duration of fixation. The reactions of various fixatives with lipids, proteins, lipoproteins, nucleic acids and carbohydrates are also discussed. The choice of fixative for analysis of tissue is restricted to those containing no elements that would be likely to contaminate the specimen.

An important property of the fixative is that it should form stable bonds which will prevent extraction of tissue components during dehydration and subsequent treatments. The fixative crosslinks proteins in the tissue both intramolecularly and intermolecularly. Crosslinks are formed both between reactive molecular groups in the tissue and between reactive groups of the tissue and the fixative itself. These reactions depend on the nature of the tissue protein and on the fixative. One important property of the fixative is that the denaturation of proteins makes them less soluble in water. Giese (1968) discusses the solubility of various proteins, lipids, carbohydrates and conjugated compounds in various solutions. Very large changes in intracellular fluid distribution during fixation have been reported by Van Harreveld and Khattab (1969). Krames and Page (1968) have shown that tissue fixed with osmium tetroxide can accumulate large amounts of calcium. The mechanism of chemical fixation of tissues is at once enormously complicated and little understood, yet a good degree of success has been achieved in analysis of biological tissue prepared by chemical fixation.

(ii) *Washing and dehydration* After primary fixation, excess fixative is removed by washing with a buffer solution before treatment with a second fixative, which is usually osmium tetroxide. The tissue is then dehydrated. Contamination of the specimen can occur at these stages since some buffers contain phosphorus, sodium, or arsenic or other elements that may well interfere with the analysis. Also $MgCl_2$ or $CaCl_2$ is sometimes included in the wash to reduce the loss of cellular materials. Osmium from the post-fixative will be widely distributed within the tissue and must be accounted for in the analysis. Dehydrating agents include alcohols, acetone, gelatin, water-miscible methacrylate resins, water-miscible epoxy resins and poly-ethylene glycol. The alcohols are by far the most frequently employed. Unfortunately, as well as being agents for replacing water in the tissue, dehydrating agents are nearly all strong organic solvents which can also cause extraction of some cell constituents. To minimise this, dehydration is performed as rapidly as possible. Millonig (1966) has suggested that ethanol causes least extraction of cellular materials but advocates the addition of NaCl to the ethanol. Feder (1960) has suggested that methanol is preferable.

Transitional solvents are sometimes used between ethanol and the embedding resin. The highly reactive propylene oxide, a commonly used transitional solvent, has been found to extract lipids, even from fixed tissue.

(iii) *Embedding* The dehydrating agent is gradually replaced by the embedding medium in order that the tissue may be completely impregnated. The major requirements of the embedding medium concerning subsequent analysis are that it should produce little extraction of cellular constituents, provide good preservation of the ultrastructure, have good stability under electron bombardment (§ 5.1.2), and cause minimum contamination of the specimen. Various embedding media have been employed including metha-crylates, polyester resins and epoxy resins. These are listed and discussed by Kushida (1965) and Glauert (1974). A number of difficulties concerning elemental analysis can arise during embedding. Many embedding media contain substantial concentrations of elements such as Cl, S, Si and Ca that may make analysis of these elements in the specimen difficult. Van Steveninck et al. (1974b) have developed a chlorine-free embedding medium for micro-analysis.

(iv) *Extraction during fixation, dehydration and embedding*. There is much evidence to show that a considerable amount of protein, lipid and carbo-hydrate is lost during fixation, dehydration and embedding of tissues. The amount of loss depends both on the type of fixative and on the methods of dehydration and embedding used. For example, Dallam (1957) has shown that 30% of phospholipid can be lost during dehydration procedures with heart and kidney, while Korn and Weisman (1966) have demonstrated that up to 90% of lipids can be lost during dehydration of amoebae after glutaraldehyde fixation even though the fixative itself causes little extraction.

For accurate analysis of elements in tissues prepared by conventional methods it is necessary to study the losses of these elements at various stages of the procedure. Methods of separating out proteins, carbohydrates and lipids from the supernatants of the solutions employed may be required so that these solutions can then be subjected to analysis by associated techniques such as spectrophotometry. The author has employed an approach using atomic absorption spectrophotometry on a simple scale to determine the loss of zinc ions from rat prostatic tissue subsequent to each preparative step. Small pieces of tissue were processed for electron microscopy and microanalysis in a fixative containing osmium tetroxide and potassium pyroantimonate. Potassium pyroantimonate (see below) was employed in a histochemical method for the detection of calcium and zinc. After each step of fixation, washing, and ethanol dehydration, the supernatants were retained for atomic absorption analysis. A control sample of the original

TABLE 4.1

Atomic absorption analysis of solutions used in the histochemical preparation (pyroantimonate method) of rat prostate tissue for subsequent analysis of zinc. Very little zinc is extracted in the preparative procedure

ppm Zinc in tissue (dry)	*Solution*	*Volume (ml)*	*ppm Zinc in solution (net)*	*Equivalent loss of zinc (ppm)*
1460	4% $KSb(OH)_6$ + 2% OsO_4 (equal vols)	2	0.01	2
	Buffer rinse	4	< 0.05	< 20
	70% Alc	4	0.01	4
	90% Alc	3	< 0.01	< 3
	90% Alc	3	< 0.01	< 3
	100% Alc	4	< 0.01	< 4
	100% Alc	4	< 0.01	< 4
1460	8% glutaraldehyde + 4% $KSb(OH)_6$ (equal vols)	2	< 0.01	< 2
	Buffer rinse	3	0.03	9
	70% Alc	4	< 0.01	< 4
	90% Alc	2	< 0.01	< 2
	90% Alc	4	< 0.01	< 4
	90% Alc	3	< 0.01	< 3
	100% Alc	3	< 0.01	< 3
	100% Alc	4	< 0.01	< 4

tissue and a sample after dehydration were also analysed. Some of the results are presented in Table 4.1. It can be seen how virtually no zinc was lost from the tissue during the fixation and dehydration steps. Thus the relatively high endogenous zinc levels were associated with cell components not leached from the tissue; also there was no loss of 'free' zinc.

Mehard and Volcani (1975) used a method of radioactive labelling to determine the loss of silicon and germanium in rat tissues during cell and organelle preparation. Conventional and freeze-substitution methods of preparation were compared. Animals were injected with ^{31}Si or ^{68}Ge and later killed. Radioactivity in the tissue samples was monitored after fixation, washing, post-fixation, dehydration and embedding, in the conventional technique; or after quenching, substitution in ether, infiltration in vinyl cyclohexane dioxide and embedding in low viscosity medium, in the freezing method. It was shown that 65–70% of ^{31}Si or ^{68}Ge was lost in liver tissue by the end of dehydration, and a further 2–12% during embedment. Other tissues varied in their elemental retention during conventional preparation. Freeze-substitution, however, was found to produce total retention of ^{68}Ge in rat tissues when monitored in this way. X-ray analysis of ultrathin sections failed to detect germanium in either preparation and was thus unable to confirm the retention of this element. However, the technique of radioactive monitoring should provide a useful way of checking tissue elemental losses. Similar studies have been performed on plant tissues prepared by freeze-substitution (Lauchli et al. 1970). Such correlative methods of analysis must be adopted if confidence is to be placed in subsequent analyses. Morgan et al. (1975) and Harvey et al. (1976) discuss these techniques further. The techniques of Pease (1966a, b) for anhydrous sectioning and staining for electron microscopy could well be extended for use with microanalysis of ultrathin sections. Unfixed tissue is prepared by freeze-drying, freeze-substitution, or dehydration with freely permeable water-soluble compounds such as ethylene glycol or glycerol to displace intracellular water and immobilise macromolecular systems. The specimen is embedded in 'prepolymerised' hydroxypropyl methacrylate with 5% divinyl benzene added as a cross-linking agent.

Lauchli et al. (1970) and Spurr (1973) describe a technique of freeze-substitution and epoxy resin embedding for relatively thick specimens (2 μm) for analysis that could be applicable to the study of ultrathin sections for the analysis of soluble electrolytes. Tissue is first quenched in liquid nitrogen-cooled isopentane. It is then transferred into precooled diethyl ether or a mixture of diethyl ether and benzamide at −80 °C and allowed to stand at that temperature for 8 days to displace the water. The tissue is then infiltrated with vinyl cyclohexane dioxide at −20 °C and vacuum embedded in a low viscosity embedding resin. Blocks are sectioned onto water, and it has been shown (Spurr 1973) that sodium is retained in the tissue. The technique is proposed as an alternative to cryo-ultramicrotomy for studying diffusible electrolytes.

Freeze-substitution has been examined as a method for preparing plant tissue for ion localisation studies (Harvey et al. 1976) and has been shown to reduce losses of sodium and potassium to just 4 %, and chlorine to only 1 %, of tissue levels.

(v) *Sectioning* General techniques for ultramicrotomy are described by Reid (1974) in an earlier volume of this series.

After polymerisation of the embedded specimen, ultrathin sections are cut with glass or diamond knives. Since the advantage of combining X-ray microanalysis with good morphology is the primary purpose of the EMMA and TEM systems, the sections must be both thin enough to yield good resolution and thick enough to have sufficient elemental content for analysis.

It is vital that in determining relative elemental concentrations across various parts of a section in the microscope that the thickness does not vary enough to cause errors in measurements. Fresh, clean and sharp knives generally aid this requirement but mass thickness measurements (§ 5.3.5b) are necessary during the analysis to check uniformity. For transmission electron microscopy sections need to be thin enough to allow electron penetration without significant absorption and to eliminate structural overlap, yet thick enough to provide enough material to cause electron scattering for image contrast. The requirements of the particular study govern the desirable thickness but this generally falls in the range 50–100 nm.

For microanalysis, there must be a sufficient number of atoms in the path of the electron beam to yield detectable X-rays. Thus the thickness required depends on the concentration of the elements to be analysed. In general section thicknesses around 100 nm are employed. Thicker (200 nm) sections can be used for analytical measurements if adjacent thinner sections are used for a complementary study of ultrastructure. Problems resulting from self absorption of X-rays with emissions in the range 1–10 keV are unlikely to be serious for sections up to thicknesses of 2 μm at 80 kV, although this absorption depends on the X-ray energy (§ 2.4.3). Methods for measuring section thickness are described by Reid (1974) and Chandler (1976).

Unfortunately, sections cut dry on the knife edge adhere very tightly to the knife, are extremely fragile and need to be flattened before they can be examined. Consequently they have to be collected on the surface of a liquid in a trough attached to the knife. This introduces another possible error of contamination or loss. Aqueous solutions of acetone or ethanol have been used as flotation agents but deionised and sterile water is most commonly used for sections for analysis. The possibility of contamination, either directly, or through acid leaching from a glass knife cannot be ruled out. Since subcellular organelles are effectively sliced open during the cutting process, any retention of elements which might have occurred within these organelles by the intact membranes in the whole tissue is destroyed and the elements may simply pass into solution in the water bath. For example, Scherrer and Gerhardt (1972) compared the location of calcium in intact spores of *Bacillus cereus*, dried onto polished carbon or aluminium discs or quartz slides and carbon coated for analysis in EPMA, with the distribution in thin sections of spores prepared after fixation, staining with uranyl acetate, dehydration and embedding. Calcium losses were measured in the supernatants after each process and found to be quite low. However, losses in the trough on the ultramicrotome knife were assumed to occur since poor X-ray signals were obtained from the thin sections, while collection of sections dry from the knife increased the calcium signal between 5 and 10 times.

The technique of flotation of sections onto ethylene glycol for electron microscopy (Pease 1966b) may well be extended for use in microanalysis where water baths have to be avoided. When unfixed tissue has been dehydrated for sectioning it is obviously necessary to avoid contact of the sections with aqueous solutions. The materials used in construction of the trough (glass, metal or plastic trough, sealing tape) must not be allowed to cause contamination. Oil films that sometimes mysteriously appear on sections cause heavy carbon contamination during exposure to the electron beam.

(vi) *Staining* Sections for electron microscopy need to be stained with materials which cause electron scattering in order to provide contrast. Staining also increases the possibility of obtaining cytochemical information about cell components. Commonly used electron stains are salts of lead (hydroxide, acetate or citrate), uranium (acetate and nitrate); stains containing phosphotungstic acid, potassium permanganate, osmium tetroxide, iron, thorium, bismuth, vanadium, indium and chromium are also used. A stain is chosen, as is the fixative, for its affinity for particular cell constituents. Lewis and Knight (1977) describe methods of staining sections for transmission electron microscopy.

Unfortunately these stains interfere with X-ray microanalysis in three major ways. The heavy metals employed are ionized by the electron beam to produce characteristic X-rays as described in Chapter 2. As shown in Table 3.2, the characteristic lines of many of the heavy metals can interfere with elements that may be required to be analysed, (see, for example § 5.3.5g).

The second major problem is that of specimen mass thickness variations. The method of quantitation proposed by Hall (1968, 1971) and outlined in Chapter 6 involves the measurement of mass thickness of the specimen as a means of determining elemental concentrations. When heavy metals are incorporated in the specimen, however, this measurement becomes confused because of the heterogenous distribution of the stain. In addition, embedding medium is of a different density to the tissue and is not evenly distributed in the specimen so that errors can arise, even in unstained tissue.

A third problem is that the presence of heavy metals vastly increases the general X-ray background and reduces the sensitivity of analysis, particularly for trace elements (§ 7.3.1).

In practice, both stained and unstained sequential sections may be examined in an attempt to correlate ultrastructure with analysis. When absolute concentrations are not required, however, the worst effect of the stain is to increase the general X-ray background and so reduce sensitivity (§ 6.2.1a). Care must be taken to allow for possible impurities within the staining complex used. This may be achieved by allowing a drop of the staining salt to dry onto a support film on a grid and then analysing for total elemental composition. In practice, it is found that lead citrate and uranyl acetate (commonly used stains) are virtually free from impurities when prepared from Analar grade reagents.

4.4.2b Cytochemical techniques

Many cytochemical methods of demonstrating or localising molecular groups are suitable for use in combination with X-ray analysis. The prerequisite is that the final reaction product must contain a detectable element. These methods are reviewed comprehensively by Pearse (1972) and by Lewis and Knight (1977). Weavers (1973), Lauchli (1975), Chandler (1975) and Bowen et al. (1976) discuss some of those most promising for X-ray analysis.

A frequently used method for the retention of some cations in soft tissues is the pyroantimonate technique. It was first devised for the demonstration of sodium but later shown to be relatively ineffective for this element, yet very suitable for a number of other cations. Each cation and each tissue demands a different method of employing the reaction, which depends on the substitution of potassium by tissue cations in the potassium pyroantimonate molecule when the salt is dissolved in the initial fixative. Various forms of the salt are available (e.g. $KSb(OH)_6$ or $K_2Sb_2O_7 \cdot 4\,H_2O$) and the reaction is strongly dependent on pH, PO_4 ions in the buffer, osmolarity, type of fixation, washing procedures, staining procedures, etc. The various methods are described by Sumi and Swanson (1971), Garfield et al. (1972), Torack and Lavalle (1970), Tandler et al. (1970) and Yarom et al. (1974).

Before attempting to fix the tissue using pyroantimonate it is recommended that a series of simple *in vitro* experiments are performed to determine the conditions of fixation required, using salt solutions of the various elements to be analysed in varying molar concentrations in place of the tissue. The resultant reaction products are analysed in precipitate form to determine the efficiency of the reaction for a certain molar concentration. Interpretation of these experimental results must be treated with some caution, however, in view of the various molecular binding forces involved for the elements within the tissue.

A method widely used with some success is as follows:

(i) Make up 100 ml of 0.01 N acetic acid and adjust to pH 8.0 with 1 N KOH. The adjustment should be done very carefully by titration or the antimony salt may not subsequently dissolve.

(ii) Heat to 80 °C for approximately $\frac{1}{2}$ hr without boiling and maintain volume at 100 ml.

(iii) Add 2 g of potassium pyroantimonate to the solution and stir with a glass rod. Leave the solution on a hot plate and continue to stir until the pyroantimonate has dissolved.

(iv) Filter off any small amount of residue.

(v) Place 3.86 g of sucrose in a beaker and pour in the antimonate solution. Stir to dissolve the sucrose and cool the solution to room temperature. This step is optional and may be omitted for some tissues.

(vi) Break a vial of 1 g of osmium tetroxide into a 100 ml pyrex bottle and add 100 ml of antimonate solution. (Alternatively 10 ml may be added to 0.1 g of OsO_4). Leave to dissolve for at least 12 hr.

(vii) Decant into a clean dropper bottle.

The solution should be kept refrigerated before use and the pH adjusted to pH 7.6 immediately before fixation (adjust pH with 1 N KOH or 0.05 N acetic acid). If the pH falls much below 7.6 spontaneous precipitation of potassium pyroantimonate occurs, and the solution becomes cloudy. This results in a much reduced precipitation of cations in the tissue.

Fixation is performed at 4°C for 1 hr without prior glutaraldehyde treatment. Some workers (Yarom and Chandler 1974) omit washing after fixation and pass the tissue directly into alcohol for dehydration. If ultrastructural preservation is not important the tissues may be stored in alcohol overnight. Other workers suggest washing in sucrose solution. The technique has been used with both perfusion and immersion methods of fixation.

The use of sucrose in the osmium fixative to adjust osmolarity is not always recommended (Glauert 1974). Osmium may be used in the unbuffered state for some tissues. If sucrose is to be used, the concentration will depend on the final osmolarity required to avoid shrinkage or swelling of organelles (Glauert 1974).

It must be remembered that subcellular precipitation of the cations with antimony depends, as do all histochemical reactions, on the availability of the element in the tissue. Elements which are strongly bound to protein may not yield reaction products during the fixation procedure. However, the element may still be in the tissue in its subcellular location, although not visible as a precipitate in the transmission image, and may be detectable. The precipitation is often extremely fine and may be missed during a cursory observation of the specimens, or if sections are subsequently stained.

It has been shown (Sumi and Swanson 1971) that the method is not suitable for retaining sodium or potassium, and that under certain conditions phosphate ions may be precipitated or may inhibit the reaction with other cations. Although the reaction is strongly sensitive to calcium, certain difficulties arise in the analysis of calcium in the presence of antimony due to their overlapping X-ray lines (§ 5.3.4g).

There are several important features of the pyroantimonate method:

(i) The precipitation is an artefact. Cations do not actually exist in discrete particulate form and will have migrated during the reaction to form the precipitates. In a tissue with much protein this migration may not be very great but again it is important to apply correlative freezing techniques to confirm elemental distribution.

(ii) Some precipitates are so fine as to be barely visible in the electron image. The operator should not abandon the technique if no precipitates appear to be present. Analysis will often indicate the presence of antimony where none is visible.

(iii) Cations will not always combine with the antimony in the same ratio. This may depend on the conditions of fixation. Thus a large precipitation may not indicate a large amount of a particular cation. This must be determined by subsequent analysis. Coarse deposits may be formed due to the migration of ions along a diffusion gradient.

(iv) Non-reproducibility of the reaction may appear to occur, especially if the fixation and embedding routine is not exactly repeated. This does not necessarily mean the cations are not retained since, for example, calcium may bind with the antimony molecule in a number of different ratios.

(v) Some non-specific precipitation may occur due to deposition of potassium or other cations (Yarom et al. 1974, 1975). This again must be determined by analysis. It has even been suggested that organic solutes in tissues may be precipitated by pyroantimonate (Winborn et al. 1972).

A non-specific precipitation of potassium pyroantimonate may be more likely when fixation is performed at a more acidic pH than 7.6. Any reduction of precipitation through washing or staining the sections may then be due to the redissolving of these precipitates, while the precipitates formed from other cations remain insoluble in the tissue. Analysis is the only way to determine the nature of the precipitates.

Sections may be lightly stained after sectioning for EM observation, although they are best analysed unstained. A suitable staining procedure which will not mask the precipitates is immersion in a lead citrate solution for 5 minutes.

The method has been shown to faithfully retain calcium and zinc in their *in vivo* subcellular concentrations in human sperm cells (Chandler and Battersby 1976b), and has been widely used, mainly for the retention of calcium in tissues (Yarom and Chandler 1974; Hales et al. 1974) but also for magnesium and manganese (Herman et al. 1971) and cadmium (Chandler and Timms 1976). The pyroantimonate method is the most widely used 'wet' method for analysis of cations in tissue. The reliability of element retention needs to be confirmed wherever possible by comparison with freezing techniques.

Although the pyroantimonate reaction is widely used it should be treated with some caution when attempting quantitative analysis of subcellular elements. A great deal more work needs to be performed to determine the factors affecting the quantity and distribution of precipitates from this reaction.

A number of other histochemical techniques exist for subcellular localisation of ions (Table 4.2) (see Lauchli 1975). Similar precautions to those described above should be taken when interpreting ion distributions by these methods.

TABLE 4.2

Precipitation reactions for demonstration of inorganic ions in biological tissues

Ion	*Precipitating agent*	*References*
(a) *Animal tissue*		
Na^+	K-pyroantimonate	Van Lennep and Komnick (1971) Komnick and Stockem (1973)
Na^+ Mg^{2+} Ca^{2+}	K-pyroantimonate	Tandler et al. (1970) Garfield et al. (1972) Yarom and Meiri (1973)
Na^+ K^+ Mg^{2+} Ca^{2+} Mn^{2+}	K-pyroantimonate	Weavers (1971)
Ca^{2+}	K-pyroantimonate	Yarom et al. (1974) Yarom et al. (1975)
Ca^{2+}	K-oxalate	Podolski et al. (1970) Yarom et al. (1975)
Zn^{2+} Cd^{2+}	K-pyroantimonate	Chandler and Timms (1976)
Cl^-	Ag-acetate or lactate	Bock (1970) Van Lennep and Komnick (1971)
Zn^{2+}	AgS	Chandler et al. (1976)
PO_4^{---}	Pb-acetate	Tandler and Solari (1969)
(b) *Plant tissue*		
Na^+	K-pyroantimonate	Levering and Thomson (1972)
Na^{2+} Mg^{2+} Ca^{2+}	K-pyroantimonate	Tandler et al. (1970)
Na^+ Ca^{2+}	Benzamide	Spurr (1972)
Ca^{2+}	NH^4-oxalate	Braatz and Komnick (1973)
Cl^-	Ag-acetate or lactate	Lauchli et al. (1974) Van Steveninck et al. (1974a) Stelzer et al. (1975)
PO_4^{---}	Pb-acetate	Tandler and Solari (1969) Libanati and Tandler (1969)
ATPases	$Pb(NO_3)_2$	Hall (1973)

4.4.2c Cryo-ultramicrotomy

Cryo-ultramicrotomy is a technique used for the retention of soluble electrolytes and cell constituents *in situ* when all forms of fixation, etc., must be avoided. Equipment is now available for cutting ultrathin frozen sections on standard ultramicrotomes (see Appendix). This work has been pioneered by such investigators as Bernhard (1965), Stumpf and Roth (1965), Christensen (1967, 1971), Hodson and Marshall (1970), Appleton (1972), and Sjoström and Thornell (1975), and is described in detail by Reid (1974). Results are very slow in forthcoming. The major difficulty lies in obtaining good enough preservation of ultrastructure for localising subcellular components for analysis. A method commonly adopted is as follows: A small piece of fresh tissue (1 mm^3) is placed on the end of a metal pin and quenched in a medium, such as isopentane, Freon 22 or propane cooled with liquid nitrogen. The tissue may then be stored in liquid nitrogen almost indefinitely. Alternatively, the tissue is frozen rapidly by pressing it against a copper block partly immersed in liquid nitrogen. The pin with frozen tissue is mounted in a cryo-ultramicrotome. The block is kept at about $-100\,°C$ and the knife at about $-80\,°C$, although lower temperatures are also used. Ultrathin (100 nm) sections are cut dry with a glass knife with no trough liquid. The transfer of sections onto suitably coated grids is often a tricky problem, and various methods have been attempted. One of the most ingenious techniques is the use of a vacuum pipette to suck the sections out horizontally from the knife edge in a ribbon (Appleton 1972), with a grid placed on the flat surface of the knife underneath. The mounted sections are freeze-dried, preferably in the cryo-chamber and under vacuum, for 2–3 hr or more. It is important to keep the sections from atmospheric moisture since they are very hygroscopic.

A coating of carbon deposited *in vacuo* would protect the sections but, in fact, many X-ray microanalytical studies have been performed on material that has not been protected in this way.

Some attempts to stain frozen, freeze-dried sections have been made, but with little success so far (Christensen 1971; Gay 1972). The method employs vapours of phosphotungstic acid or osmium tetroxide and is difficult to control.

Ultrathin frozen sections have been examined in the transmission electron microscope at room temperatures, while in some techniques being developed at the present time sections are analysed in the frozen state in either the SEM or the STEM (Echlin and Moreton 1974; Saubermann and Echlin

1975). This may be done before or after freeze-drying. Analysis of sections while they are maintained at very low temperatures may help to offset the thermal effects of electron bombardment (Hall and Gupta 1974). The argument for analysing frozen, ultrathin, hydrated sections is that with the frozen vitreous water still present in the tissue, migration of ions will have been reduced.

The parameters governing the quality of frozen sections are: freezing method; temperature of block and knife; cutting speed; angle (and sharpness) of knife; efficiency of freeze-drying or of maintaining hydration; efficiency of collection and flattening of sections; transfer; storage.

Many recent investigations have been concerned with developing methods of very rapid specimen cooling. This is in order to reduce ice crystal formation in the tissue (Moor 1973) which interferes with the electron image and possibly redistributes dissolved ions. The temperature of the tissue block during cutting is important. If thawing (Hodson and Marshall 1972) does occur during cutting then ionic movement might take place. Thus, on the one hand, the block needs to be kept as cold as possible. On the other hand, it becomes extremely difficult to cut tissue at very low temperatures ($-120\,^{\circ}$C), because of increased brittleness. Glass knives with a range of cutting angles have been used routinely for conventional microtomy. It may be that superior sections will be cut of frozen material using knives with very small cutting angles to overcome the difficulties of brittleness, ripple and cleaving, or alternatively with diamond knives.

The transfer of the specimen from the cryo-ultramicrotome to the microscope also presents serious problems. Devices are now being developed to allow the transfer to occur without the deposition of ice from atmospheric vapour (Echlin and Moreton 1974; Saubermann and Echlin 1975).

4.4.2d Replication

The analysis of particles in biological tissue (e.g. asbestos in lung tissue) and of cell fractions may be performed by the method of extraction replication described by Henderson and Griffiths (1972). In this technique a thin plastic mould of the surface of the specimen is formed and then stripped off. Particles originally in the specimen surface are removed with the replica. The replica is coated with carbon, and the plastic is then dissolved away. The carbon replica and the particles are retained and examined in the microscope.

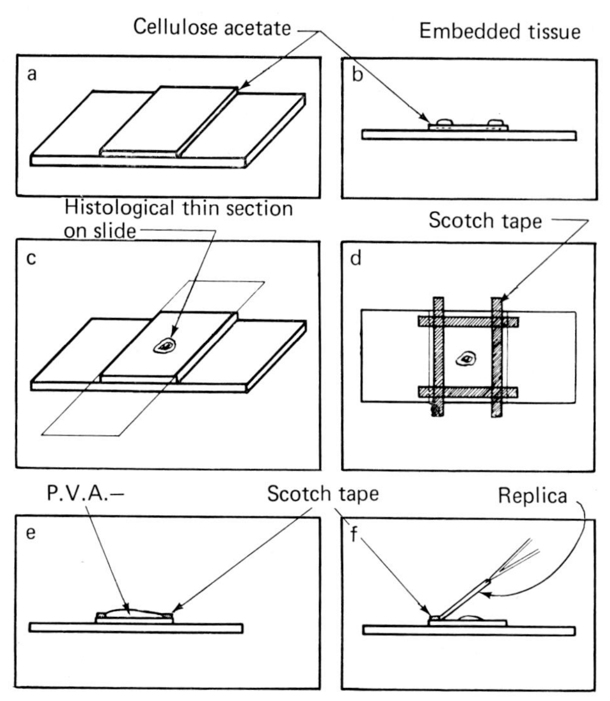

Fig. 4.2. Preparation of an extraction replica from biological tissue. Soft tissue (b) or histological sections (c) are embedded in the softened cellulose acetate. Polyvinyl alcohol (PVA) is poured into a well (d), allowed to harden and removed (f).

The method is as follows:

(i) The tissue to be examined is embedded in acetone-softened sheets of cellulose acetate (0.5 mm thick, $25 \times 25\ mm^2$) and mounted on glass microscope slides for ease of handling (Fig 4.2). Fresh unfixed tissue can be embedded directly, whereas tissue which is softer and moist can be dehydrated *in vacuo* to ensure sufficient rigidity. Paraffin is removed from standard histological specimens by immersion in xylene before ethanol dehydration and embedding in the cellulose acetate. Care is necessary to prevent the tissue from being completely immersed in the cellulose acetate.

(ii) The acetate is allowed to harden for about an hour.

(iii) The tissue in the acetate is outlined with strips of adhesive tape in order to form a shallow well and to aid removal of the replica.

(iv) A 10% (v/v) solution of polyvinyl alcohol (PVA) is applied to cover the tissue. A 5% solution can be applied to smooth surfaces.

(v) The PVA is allowed to harden overnight at room temperature or for 1 to 2 hr on a hot-plate. Heating may, however, introduce artefacts through the inclusion of air bubbles in the replica.
(vi) On hardening, the PVA replica of the tissue surface is stripped off; successive strippings through the embedded tissue can also be made. (Small particles or foreign matter embedded in the tissue surface will be extracted by the PVA film during the stripping procedure).
(vii) The PVA replica is reversed and coated with a film of carbon *in vacuo* to a thickness of about 20 nm.
(viii) The PVA is removed from the carbon replica film by floating the replica in a hot water bath at approximately 97 °C for 15 min. The particles will be left attached to the floating carbon replica.
(ix) The replica is mounted on an electron microscope grid and is ready for analysis.

The method has been used for the study of fibrous particles in lung tumours (Henderson et al. 1970), for platy particles in ovary and cervix, and for various minerals extracted from histological sections of stomach tumours (Henderson et al. 1973).

4.4.2e Ashing

It is often necessary to examine mineral particles or inorganic matter in an organic matrix where the method of replication described in § 4.4.2d is not possible. A frequently used technique, for example for the study of asbestos particles in lung tissue, is to remove the organic matrix by a process of ashing or incineration. It is important to determine, by preliminary control experiments, that the method of ashing does not affect the chemical composition of the particles to be examined.

One of the simplest and most effective methods of preparation is to simply mount a section of organic tissue containing the particles on a suitable substrate (e.g. beryllium, carbon or quartz) and incinerate it in an oxygen flow oven for a number of hours at about 500 °C. The ashed residue may be carbon coated before analysis in the SEM. The specimen may be in the form of a thick section which may be examined with a light microscope before the ashing procedure to provide details of morphology. The section is cleared of paraffin wax with xylene before ashing.

An alternative to high-temperature ashing is the method of oxygen plasma ashing which takes place at less than 100 °C (Thomas and Hollahan 1974). The sample (e.g. soft tissue) is mounted, as before, on a suitable stub or neutral substrate and placed inside the reaction chamber of the ashing device around which a plasma excitation coil electrode is wound. Oxygen is passed over the specimen at a controlled pressure (0.5 to 1.0 Torr) and a radio-frequency excitation is generated in the coil. Over a period ranging up to a few hours the organic matter is removed by the action of highly excited plasma species from the gas which react chemically with the specimen surface and produce volatile compounds which are removed by the vacuum. The method is most suitable for tissue fragments less than 0.5 g in weight. Microincineration may also be employed for TEM studies when the original section is too thick to be examined by transmission, or even with ultrathin sections mounted on carbon-coated grids. The ashing procedure leaves the specimen as an inorganic skeleton thin enough to be studied in the TEM. Methods of micro-incineration using oxygen plasma ashing are described by Thomas (1974).

When the ashing procedure is employed to simply extract particles from tissue for TEM studies, the residue must be subsequently handled in such a way as to avoid contaminating it with dust and other spurious airborne particles. If the ashing takes place in a crucible or boat (for example of platinum), the residue is carefully washed with distilled water, transferred to a clean centrifuge tube, and centrifuged for 15 mins at 1000 g. The bulk of the supernatant is discarded leaving the residue suspended in a few microlitres of water at the base of the tube. A micropipette is then used (100 μl size) to deposit a droplet on a carbon-coated microscope grid. The droplet is allowed to dry under a 40 W lamp or in an oven under a Petri dish to prevent dust contamination. The grid is then ready to be carbon coated in *vacuo* and examined in the TEM. Alternatively, the droplet may be allowed to dry onto a suitable support (carbon, beryllium, etc) and carbon coated for SEM studies.

4.5 Standards

Standards are used in either bulk form (where the electron beam is totally stopped) or in a thin form where the electron beam penetrates the specimen. The thin type of standard may be mounted on an opaque substrate or on a grid or holder allowing transmission of the electron beam. With thick specimens it is necessary to take into account the effects of X-ray absorption and fluorescence and of atomic number (see Chapter 6). An ideal standard allows these factors to be minimised when making direct comparisons with the specimen.

Although many quantitative analyses have been performed using standards which do not closely resemble the composition of the specimen, it is necessary in some applications, especially biological ones, to prepare standards of a similar nature to the specimen under investigation. In particular it may be important to achieve a similar surface texture, thickness, and internal homogeneity. The standard and specimen must both be stable in the electron beam.

A standard specimen may be simply one containing a substantial quantity of the element or elements to be examined for the purpose of calibrating the X-ray detector system. Or it may contain a known concentration of each element for making suitable comparisons with the specimen under investigation so that quantitative analyses may be performed (see Chapter 6).

For simple calibration purposes, e.g. for tuning the X-ray crystal spectrometer to the correct wavelength for the element concerned, the essential requirement is that the standard should have a smooth even surface so that its position in the specimen holder can be accurately determined. When a wavelength crystal spectrometer is to be used, the height of the surface of the standard must always be exactly the same as that of the specimen to be subsequently examined or accurate focusing of the X-ray beam by the spectrometer will be impaired (§ 5.3.3). In most transmission instruments the specimen is held against a fixed surface or in a fixed position within the specimen stage. With SEM instruments, however, the specimen height may have to be adjusted with the stage manipulators. The position of the specimen surface should be accurate to at least 25 μm and the surface smooth to within 1 μm. The specimen height is not so critical when a SSD is employed since no X-ray focusing is involved. The effect of an irregular surface on absorption of the X-ray signal is illustrated in Fig 4.1.

4.5.1 *Qualitative standards*

For simple qualitative work a standard need only contain the element of interest so that the X-ray detector may be calibrated for energy (or wavelength). To prepare a simple standard of almost any element, a salt or compound is chosen which contains the element (for example, for calcium some calcium sulphate, chloride, or carbonate may be used). A suspension of the powder (or a solution if the salt is soluble in water) is prepared and a droplet (10 μl) is then allowed to dry on a grid previously coated with a support film, such as carbon or formvar, or, for SEM studies, on a neutral support, such as carbon. The suspension or solution will form small particles or crystals that will adhere to the support film and may be used as a simple standard for analysis. Other examples are NaCl for Na and Cl; $MgSO_4$ for Mg and S; $MgSiO_3$ (talc) for Mg and Si; $CaPO_4$ for Ca and P; etc. The specimens may need to be carbon coated to stabilise them in the beam. Pure metal standards may also be prepared from a number of elements such as Mg, Al, Mn, Fe, Ti, etc., by punching out discs of the appropriate size from sheet metal. Discs may be prepared by turning a rod of the metal on a lathe to the correct diameter, and then cutting, polishing, and washing clean. Diamond paste abrasive is often used for polishing (Goodhew 1972).

4.5.2 *Quantitative standards*

As discussed in Chapter 6, standards may be required to calibrate the X-ray signal so that elemental ratios may be determined or absolute concentrations calculated. Standards are often of a type or composition similar to that of the specimen being examined, but this is not necessarily so. Mineralogical and metallurgical standards differ from those required for biological work and each is discussed separately.

4.5.2a *Metallurgical and mineralogical standards*

For analysis of bulk material, pure elements are often used for standards, and are also in the bulk form. During such an analysis the observed characteristic X-ray intensity from an element in the sample is compared, under identical operating conditions, with the characteristic X-ray intensity from the same element in an appropriate bulk standard (§ 6.1). Quantitative X-ray microanalysis of bulk specimens in the EPMA or SEM is outside the scope of this book. The reader is referred to the list of Further Reading at the end of Chapter 6.

In a specimen which is thin enough to be examined by transmission electron microscopy such bulk standards are inadequate because the emitted X-ray intensity from a specimen is a function of its thickness as well as its elemental composition. Quantitative analyses of thin metal foils using bulk standards have been performed (§ 6.2.2c), but lengthy correction procedures (Nasir 1972) are usually necessary for absorption and fluorescence effects.

Cliff and Lorimer (1972) have used a technique for thin metal foils in which the specimen acts as its own standard, providing that the bulk analysis of the specimen is known from chemical tests. The method allows the analysis of small regions of the specimen to be made for calculation of chemical mass fractions. The specimen is in the form of an electropolished thin metal foil for TEM studies (see § 6.2.2).

When a knowledge of the ratio of two elements in individual particles in a powder dispersion, or in a submicron region of a metal foil, is required, the standard is a specimen (also a powder dispersion) for which the bulk analysis is well known. The bulk analysis is provided by chemical analysis, atomic absorption analysis, or X-ray fluorescence analysis. Quantitative analysis of individual particles is then performed after analysis in the microscope of a large area of the specimen for which the bulk analysis applies. The technique of Rowse et al. (1974) for preparing such a standard for the analysis of Si : Al ratios is described here. Its application to quantitative analysis is discussed in § 6.2.2. Other standards for different elemental ratios may be made in a similar way.

(i) Prepare a slurry of silica (SiO_2) in distilled water, about 10% by wt.
(ii) Prepare a slurry of gibbsite ($Al(OH)_3$) in distilled water, about 10% by wt.
(iii) Add a deflocculent, such as sodium polyacrylate, to each slurry, if necessary.
(iv) Hold each suspension in a test-tube in an ultrasonic bath for 1 min to break down aggregates.
(v) Mix the two slurries together in the required proportions (to simulate the expected Al : Si ratio in the sample), and stir vigorously.
(vi) Filter the mixture by suction through a large area of filter paper.
(vii) Wash the slurry with acetone and heat to 110°C to remove solvent and water.
(viii) Take one half of the slurry for chemical analysis.
(ix) Embed the other half of the slurry in an epoxy resin by impregnation under vacuum and section at about 100–120 nm thickness with a diamond knife.
(x) Mount sections on a carbon or formvar film on an EM grid.
(xi) If necessary, coat with a thin layer of carbon.

Standards of other composite minerals can be prepared in a similar way. Alternatively a suspension of the slurry in water or ethanol may be dried down onto a carbon-coated EM grid.

The X-ray detection system is calibrated as described in § 6.2.2. For calibration of the detector's relative collection efficiency of X-rays from different elements (X-rays with different energies) a number of composite elemental standards are employed. The calibration method is discussed in § 6.3. The choice of suitable composite mineral standards for thin specimens is discussed by Sweatman and Long (1969) and by Cliff and Lorimer (1975). In general such standards of minerals are prepared from powdered samples as already described (§ 4.3.2). Cliff and Lorimer (1975) have used a series of amphiboles (hornblende, actinolite, grunerite, tremolite) which are stable in the electron beam and for which accurate chemical analyses are available.

For the calculation of absolute mass fractions in thin specimens, in which the ratio of characteristic to white radiation is measured (Hall 1971) (see Chapter 6) the standard must be thin and need only contain a homogeneous dispersion of the element of interest in a known proportion. Sweatman and Long (1969) discuss the importance of accurately determining the composition of standards by alternative techniques and the need to ensure the stability of standards in the electron beam.

Philibert et al. (1970) used thin evaporated metal films as absolute standards but their technique depends on a knowledge of the mass thickness of the standard. Thicknesses of ultrathin metal films cannot easily be measured in the average laboratory and such standards are available only from specialised laboratories. When the thickness of a metal film or foil approaches 1 μm then the parallax method of Lorimer et al. (1975) (§ 4.2.3) may be used to measure it.

4.5.2b Biological standards

Although inorganic standards are widely used for quantitative analysis of biological specimens, since they have well-defined composition and stability in the electron beam, there are certain disadvantages in using them in bulk form for calibration of thin specimens. These disadvantages include the frequent necessity to correct for absorption and fluorescence of X-rays. Thin specimens of inorganic material can be used as standards to avoid these difficulties (Hall 1968). The minerals sandine and albite have been used by Hall and Peters (1974) and Hall and Gupta (1974), while thin sections of dentine have been used for the determination of P and Ca in blood platelets (Skaer et al. 1974). Johnson (1969) used the mineral fluoroapatite for determining Sr and Ca in bone, and Yarom et al. (1973) prepared vaporized gold standards on carbon-coated electron microscope grids for determination of intra-articular gold. Apatite ($3Ca_3(PO_4)_2 \cdot Ca(OH)_2$) is a convenient mineral standard with which to perform quantitative studies on organic specimens and has been used by Yarom et al. (1974) to determine calcium in heart muscle. Apatite has also been widely used as a standard for analysis of calcified tissues (Höhling and Nicholson 1975).

Spurr (1975), however, found discrepancies of up to 23% in concentrations of sodium in botanical specimens when comparing data using organic and inorganic standards. Because the mean atomic density of organic and inorganic specimens differs, the generation of white radiation by the electron beam is not the same. This creates difficulties when using the continuum method for quantitation (§ 6.2.1) even for thin specimens. Consequently many workers prefer to prepare standards of a similar composition to the specimen under investigation. Whatever standard is chosen it is essential to ensure that it is stable in the electron beam. Hall and Peters (1974) discuss the comparison of organic and mineral standards for quantitation.

To simulate biological tissue, suitable organic standards can be prepared by adding salts of the required elements to organic material and cutting sections. Some simple methods of preparing standards are described below:-

(i) *Albumin matrix (SEM studies)* Ingram and Hogben (1968) used a 20% solution of bovine serum albumin mixed with 5% glycerin. Known quantities of given salts containing the element of interest (e.g. NaCl or KNO_3) are added in solution. Drops of the solution are quickly frozen in liquid nitrogen, and freeze dried. They may then be fixed in OsO_4 vapour or left unfixed according to the treatment of the specimen being examined. The freeze-dried material is then embedded in the same resin as the specimen, sectioned with a steel knife and mounted on a carbon-coated quartz slide. The standard is then coated with carbon (20 nm) in a vacuum evaporator.

(ii) *Tissue homogenate matrix (SEM studies)* Various concentrations of electrolytes (such as NaCl, KCl) are added to kidney homogenates to provide a regular distribution. The homogenate is then frozen and sectioned, freeze dried and analysed. The elements are found to be dispersed evenly in a biological matrix which closely resembles that of the specimen (Kriz et al. 1972).

(iii) *Gelatin matrix (SEM studies)* Maroudas (1972) prepared standards of $CaSO_4$ in gelatin for the study of Ca in articular cartilage, the standard having an elemental composition close to that of the cartilage matrix.

A 1% gelatin solution and a 0.1% $CaSO_4$ solution (Analar $CaSO_4 \cdot 2H_2O$) are prepared. The two solutions are mixed in different proportions to give various concentrations of $CaSO_4$. Calculated volumes of each concentration are spread onto quartz slides to give films of different thickness. The films are freeze-dried and carbon-coated prior to examination in the electron probe.

(iv) *Epoxy resin matrix (TEM studies)* In the method of Chou et al. (1973), a small quantity of the element in the form of a salt or oxide (e.g. CuO) is put in a drop of molten agar and mixed. The agar is oven-dried to a thin film. It is then dehydrated in an ethanol series and embedded in Epon in the usual way. 100 nm sections are cut with a diamond knife and placed on formvar or carbon-coated grids. The sections are then ready for analysis with a defocused illuminating beam to cover a large number of particles in the section.

A similar technique has been used by Hall and Peters (1974) for the analysis of sodium using sections of various minerals in embedding medium. The choice of minerals is important from the point of view of stability in the electron beam. Elements may also be added in known quantities to embedding material and sections prepared for comparison with the specimen.

(v) *Macrocyclic polyether salt complexes in epoxy resin (SEM or TEM studies)* Spurr (1974) describes a method of introducing sodium or potassium into low viscosity epoxy resins for thick or thin specimens. The method involves the incorporation of electrolytes (or other elements) into a cyclic polyether which is then mixed into the epoxy resin in the appropriate proportions. The resin is polymerised as usual and sectioned. The sodium or potassium is in the form of a thiocyanate. Chandler (1976) describes a similar technique for the introduction of Na, K or Zn into Araldite epoxy resin in concentrations from 0.01% to 1% using the *cyanide* salts. The method for potassium is as follows:

(1) Add stoichiometric quantities of anhydrous KCN (MW 65.12) and dichlorohexyl (18) crown-6 (MW 372.5) (Aldrich Chemicals UK; see Appendix) in the proportions 0.065 : 0.372 and dissolve by stirring in 10 ml of methyl alcohol.
(2) Leave the mixture overnight, in a fume cupboard, to dry. The resulting mix is a semicrystalline honey-coloured viscous mass.
(3) Make up 20 ml of embedding medium.
(4) To the KCN and crown-6 mixture (0.435 g) add 1.515 g of Araldite to produce a 2% potassium mixture.
(5) Dilute this mixture with Araldite to obtain the appropriate concentrations of potassium in a percentage range down to 0.01%.
(6) Polymerise the resin, in gelatin capsules, at 60 °C for 36 hr.
(7) Cut thick or ultrathin sections and mount them on support films. The sections should be carbon coated before analysis.

TABLE 4.3

Standards for biological microanalysis

Type	*Specimen*	*Materials*	*Preparation*
1. Inorganic	Thick/thin	Pure metals, formulated glasses, minerals, salts	Various
2. Organic matrix salt mixture	Thin	Albumen or gelatin, agar–agar or mixtures, salts	Shock freeze, freeze-dry, mixtures treated like thin sections
3. Pellets	Bulk	Organic salts, organometallics	Compressed into pellets or into standard holders
4. Macrocyclic polyether salt complexes in epoxy resins		Polyethers, anhydrous salts, epoxy resin	Dissolve macrocyclic polyether salt complex in epoxy resin and polymerise
5. Tissue homogenates	Thick	Tissues (kidney etc), salts	Homogenates and salts, treated like tissue
6. Salt solutions	Bulk	Aqueous solutions of electrolytes	Shock freeze to obtain droplets
7. Substrates infused with salts	Bulk	Filter paper/membrane, salts	Metered droplets of salt solution on filter paper, shock freeze and freeze dry
8. Single crystals	Thick/thin	Organic salts and organometallic salts	Minute crystals measured by optical microscopy, carbon coated
9. Microdroplets	Bulk	Pico- to nanolitre droplets, salts	Dispense droplets on holders, various methods to ensure uniform spot thickness
10. Thin films	Thin	Solutions of alkali halides in ethanol and colloidon	Metered drops on C coated coverslip and evaporate by ethanol
11. Thin metallic layers	Thin	Metals on flat substrates	Vac evaporation of metals on substrate
12. Cl epoxy resins	Thin	Resin containing Cl	As normally for thin specimens

Examination of a range of potassium concentrations from 0.01% to 1% (Chandler 1976) has indicated a correlation coefficient of linearity between concentration and the ratio of characteristic to white radiation of 0.9985. The construction of a calibration curve using these standards is described in § 6.3.1.

Other standards employed for quantitation of biological specimens are listed in Table 4.3 and are discussed by Spurr (1975). Many are described in the book edited by Hall et al. 1974.

References

Andersen, C. A. (1967), An introduction to the electron probe microanalyser and its application to biochemistry, in: Methods of biochemical analysis, Vol. 15, D. Glick ed. (Wiley Interscience, New York), p. 148.

Appleton, T. C. (1972), Dry ultrathin frozen sections for electron microscopy and X-ray microanalysis: the cryostat approach, Micron *3*, 81.

Bearden, J. A. (1964), X-ray wavelengths, NYO-10586, U.S. Atomic Energy Commission, Oak Ridge, Tennessee, USA.

Bernhard, W. (1965), Ultramicrotomie à basse température, Annls Biol. *4*, 5.

Bock, P. (1970), Elektronenmikroskopischer Nachweis von Na^{++}, Ca^{++} und Cl^{-} in der lactierenden Milchdruse des Meerschweinchens, Cytobiologie *2*, 68.

Bowen, I. D., T. A. Ryder and N. L. Downing (1976), An X-ray microanalytical azo dye technique for the localisation of acid phosphatase activity, Histochemistry *49*, 43.

Boyde, A. and V. R. Switsur (1963), Problems associated with the preparation of biological specimens for microanalysis, in: X-ray optics and X-ray microanalysis, H. Pattee, V. E. Cosslett and A. Engstrom, eds (Academic Press, New York), p. 499.

Braatz, R. and H. Komnick (1973), Vacuolar calcium segregation in relaxed myxomycete protoplasm as revealed by combined electrolyte histochemistry and energy dispersive analysis of X-rays, Cytobiologie *8*, 158.

Bradley, D. E. (1965), The preparation of specimen support films, in: Techniques for electron microscopy, 2nd edn., D. H. Kay, ed. (Blackwell, Oxford) p. 58.

Chandler, J. A. (1975), Electron probe X-ray microanalysis in cytochemistry, in: Techniques of biochemical and biophysical morphology, Vol. 2, D. Glick and R. Rosenbaum, eds (Wiley Interscience, New York), p. 308.

Chandler, J. A. (1976), A method for preparing absolute standards for quantitative calibration and measurement of section thickness with X-ray microanalysis of biological ultrathin specimens in EMMA, J. Microscopy *106*, 291.

Chandler, J. A. and S. Battersby (1976a), X-ray microanalysis of ultrathin frozen and freeze dried sections of human sperm cells, J. Microscopy *107*, 55.

Chandler, J. A. and S. Battersby (1976b), X-ray microanalysis of zinc and calcium in ultrathin sections of human sperm cells using the pyroantimonate technique, J. Histochem. Cytochem. *24*, 740.

Chandler, J. A. and M. S. Morton (1976), The determination of elemental spatial concentration of thin specimens using X-ray microanalysis and atomic absorption spectrophotometry, Analyt. Chem. *48*, 1316.

Chandler, J. A. and B. G. Timms (1976), The effect of testosterone and cadmium on the rat lateral prostate *in vitro*, J. Endocrinol. *69*, 22 P.

Chandler, J. A., B. G. Timms and M. S. Morton (1977), Subcellular distribution of zinc in rat prostate studied by X-ray microanalysis, I. Normal prostate. Histochem. J. *9*, 103.

Chou, C. K., J. A. Chandler and R. D. Preston (1973), Microdistribution of metal elements in wood impregnated with a copper-chrome-arsenic preservative as determined by analytical electron microscopy, Wood Sci. Technol. *7*, 151.

Christensen, A. K. (1967), A simple way to cut frozen thin sections of tissue at liquid nitrogen temperature, Anat. Rec. *157*, 227.

Christensen, A. K. (1971), Frozen thin sections of fresh tissue for electron microscopy with a description of pancreas and liver, J. Cell Biol. *51*, 772.

Cliff, G. and G. W. Lorimer (1972), Quantitative analysis of thin foils using EMMA-4 – the ratio technique, Proc. 5th Eur. Conf. Electron Microscopy, Manchester, p. 140.

Cliff, G. and G. W. Lorimer (1975), The quantitative analysis of thin specimens, J. Microscopy *103*, 203.

Comer, J. J. (1971), Specimen preparation, in: The electron optical investigation of clays, J. Gard, ed. (Mineralogical Society, London).

Dallam, R. D. (1957), Determination of protein and lipid lost during osmic acid fixation of tissues and cellular particulates, J. Histochem. Cytochem. *5*, 178.

Drury, R. A. B. and E. A. Wallington (1967), Carleton's histological technique, 4th edn. (Oxford University Press).

Duncumb, P. (1962), Enhanced X-ray emission from extinction contours in a single crystal gold film, Phil. Mag. *7*, 2101.

Echlin, P. (1974), Coating techniques for scanning electron microscopy, in: Proc. 7th SEM Symp., O. Johari, ed. (IITRI, Chicago), p. 1019.

Echlin, P. and R. Moreton (1974), The preparation of biological materials for X-ray microanalysis, in: Microprobe analysis as applied to cells and tissues, T. Hall, P. Echlin and R. Kaufmann, eds. (Academic Press, London).

Feder, N. (1960), Some modifications in conventional techniques of tissue preparation, J. Histochem. Cytochem. *8*, 309.

Garfield, R. E., R. M. Henderson and E. E. Daniel (1972), Evaluation of the pyroantimonate technique for localisation of tissue sodium, Tissue and Cell *4*, 575.

Gay, J. L. (1972), X-ray microanalysis in the development of oospores of the fungus *saprolegnia*, Micron *3*, 139.

Gehring, K., A. Doerge, W. Nagel and K. Thurau (1971), The use of scanning electron-microscope in connection with a solid state detector for analysis and localisation of electrolytes in biological tissue, Proc. Int. Symp. on Modern Technology in Physiological Sciences, Munich.

Giese, A. C. (1968), Cell physiology, 3rd Edn., (W. B. Saunders & Co., USA).

Glauert, A. M. (1974), Fixation, dehydration and embedding of biological specimens, in: Practical methods in electron microscopy, Vol. 3, A. M. Glauert, ed. (North-Holland, Amsterdam).

Goodhew, P. J. (1972), Specimen preparation in materials science, in: Practical methods in electron microscopy, Vol. 1, A. M. Glauert, ed. (North-Holland, Amsterdam).

Gullasch, J. and R. Kaufmann (1974), Energy dispersive X-ray microanalysis in soft biological tissue: relevance and reproducibility of the results as depending on specimen preparation (air drying, cryofixation, cool-stage techniques), in: Microprobe analysis as applied to cells and tissues, T. Hall, P. Echlin and R. Kaufmann, eds (Academic Press, New York).

Hales, C. N., J. P. Luzio, J. A. Chandler and L. Herman (1974), Localisation of calcium in the smooth endoplasmic reticulum of rat isolated fat cells, J. Cell Sci. *15*, 1.

Hall, J. L. (1973), Enzyme localisation and ion transport, in: Ion transport in plants, W. P. Anderson, ed. (Academic Press, London and New York), p. 11.

Hall, T. A. (1968), Some aspects of the microprobe analysis of biological specimens, in:

Quantitative electron probe microanalysis, K. F. J. Heinrich, ed. (NBS Special Publication 298, Washington, DC).

Hall, T. A. (1971), The microprobe assay of chemical elements, in: Physical techniques in biochemical research, Vol. 1A, 2nd edn. G. Oster, ed. (Academic Press, New York).

Hall, T. A., P. Echlin and R. Kaufmann, eds (1974), Microprobe analysis as applied to cells and tissues, (Academic Press, London and New York).

Hall, T. A. and B. L. Gupta (1974), Measurement of mass loss in biological specimens under an electron microbeam, in: Microprobe analysis as applied to cells and tissues, T. A. Hall, P. Echlin and R. Kaufmann, eds (Academic Press, London and New York).

Hall, T. A. and H. J. Höhling (1969), The application of microprobe analysis to biology, in: X-ray optics and microanalysis, G. Mollenstadt and K. H. Gaukler, eds (Springer, New York), p. 582.

Hall, T. A. and P. D. Peters (1974), Quantitative analysis of thin sections and the choice of standards, in: Microprobe analysis as applied to cells and tissues, T. Hall, P. Echlin and R. Kaufmann, eds (Academic Press, New York), p. 229.

Hall, T. A., P. D. Peters and M. C. Scripps (1974), Recent microprobe studies with an EMMA-4 analytical microscope, in: Microprobe analysis as applied to cells and tissues, T. Hall, P. Echlin and R. Kaufmann, eds (Academic Press, London and New York).

Hall, T. A., H. D. E. Rochert and R. L. de C. H. Saunders (1972), X-ray microscopy in clinical and experimental medicine, (Charles Thomas, Illinois).

Harvey, D. M. R., J. L. Hall and T. J. Flowers (1976), The use of freeze-substitution in the preparation of plant tissue for ion localization studies, J. Microscopy *107*, 189.

Henderson, W. J., J. A. Chandler, G. Blundell, C. Griffiths and J. Davies (1973), The application of analytical electron microscopy to the study of diseased biological tissue, J. Microscopy *99*, 183.

Henderson, W. J., J. Gough and J. Harse (1970), Identification of mineral particles in pneumoconiotic lungs, J. clin. Path. *23*, 104.

Henderson, W. J. and K. Griffiths (1972), Shadow casting and replication, in: Principles and techniques of electron microscopy, Vol. 2, M. A. Hayat, ed. (Van Nostrand Reinhold, New York).

Herman, L., T. Sato and B. A. Weavers (1971), An investigation of the pyroantimonate reaction for sodium localisation using the analytical electron microscope, EMMA-4, Proc. 29th Ann. Meeting EMSA.

Hodson, S. and J. Marshall (1970), Ultracryotomy: a technique for cutting ultrathin sections of unfixed frozen biological tissue for electron microscopy, J. Microscopy *91*, 105.

Hodson, S. and J. Marshall (1972), Evidence against through-section thawing whilst cutting on the ultracryotome, J. Microscopy *95*, 459.

Höhling, H. J. and W. A. P. Nicholson (1975), Electron microprobe analysis in hard tissue research, J. Microscopie Biol. Cell. *22*, 185.

Holland, L. (1956), Vacuum deposition of thin films (Chapman & Hall, London.)

Hutchins, G. A. (1966), Thickness determination of thin films by electron probe microanalysis, in: The electron microprobe, T. D. McKinley, K. F. J. Heinrich and D. B. Wittry, eds (Wiley, New York), p. 390.

Hutchins, G. A. (1969), Surface analysis with the electron probe, in: Developments in applied spectroscopy, Vol. 7A, E. L. Grove and A. J. Perkins, eds (Plenum Press, New York), p. 325.

Ingram, F. D. and C. A. M. Hogben (1968), Procedures for the study of biological soft tissue with the electron microprobe, in: Developments in applied spectroscopy, Vol. 6, W. K. Baer, H. J. Perkins and E. L. Grove, eds (Plenum Press, New York), p. 43.

Ingram, F. D., M. J. Ingram and C. A. M. Hogben (1972), Quantitative electron probe analysis of soft biological tissue for electrolytes, J. Histochem. Cytochem. *20*, 716.

Jacobs, M. H. and J. Baborovska (1972), Quantitative microanalysis of thin foils with a combined electron microscope – microanalyser (EMMA-3), Proc. 5th Eur. Conf. Electron Microscopy, Manchester, p. 136.

Johnson, A. R. (1969), The distribution of strontium in the rat femur as determined by electron microprobe analysis, in: Proc. 4th Nat. Conf. on Electron Probe Analysis, article no. 38, Pasadena, California.

Komnick, H. and W. Stockem (1973), The porous plates of coniform chloride cells in mayfly larvae: high resolution analysis and demonstration of solute pathways, J. Cell Sci. *12*, 665.

Korn, E. D. and R. A. Weisman (1966), Loss of lipids during preparation of amoeba for electron microscopy, Biochem. biophys. Acta *116*, 309.

Krames, B. and E. Page (1968), Effect of electron microscope fixatives on cell membranes of the perfused rat heart, Biochem. biophys. Acta *150*, 24.

Kriz, W., J. Schnermann, H. J. Höhling, A. P. Von Rosenstiel and T. A. Hall (1972), Electron probe microanalysis of electrolytes in kidney cells. Problems and results, in: Recent advances in renal physiology, H. Wirz and F. Spinelli, eds (Karger, Basel), p. 162.

Kushida, H. (1965), Dehydration and embedding for electron microscopy, II. Embedding, J. Electron Microscopy *14*, 251.

Lauchli, A. (1967), Zur Technik der Herstellung biologischer Preparate für die histochemische Untersuchung mit der Röntgen-Mikrosonde, Histochemie *11*, 286.

Lauchli, A. (1975), Precipitation technique for diffusible substances, J. Microscopie Biol. Cell. *22*, 239.

Lauchli, A., A. R. Spurr and R. W. Wittkopp (1970), Electron probe analysis of freeze substituted epoxy resin embedded tissue for ion transport studies in plants, Planta *95*, 341.

Lauchli, A., R. Stelzer, R. Guggenheim and L. Henning (1974), Precipitation techniques as a means for intracellular ion localisation by use of electron probe analysis, in: Microprobe analysis as applied to cells and tissues, T. Hall, P. Echlin and R. Kaufmann, eds (Academic Press, New York and London), p. 107.

Levering, C. A. and W. W. Thomson (1972), Studies on the ultrastructure and mechanism of secretion of the salt gland of the grass spartina, Proc. 30th Ann. Meeting EMSA.

Lewis, P. R. and D. P. Knight (1977), Staining methods for sectioned material, in: Practical methods in electron microscopy, A. M. Glauert, ed. (North-Holland, Amsterdam).

Libanati, C. M. and C. J. Tandler (1969), The distribution of the water soluble inorganic orthophosphate ions within the cell: accumulation in the nucleus, J. Cell Biol. *42*, 754.

Lorimer, G. W., G. Cliff and J. Clarke (1975), Determination of the thickness and spatial resolution for the quantitative analysis of thin foils, Institute of Physics meeting on Electron Microscopy and Analysis, Bristol 1975.

Maroudas, A. (1972), X-ray microprobe analysis of articular cartilage, Connective Tissue Res. *1*, 153.

Marshall, D. J. and T. A. Hall (1968), Electron probe X-ray microanalysis of thin films, Br. J. app. Phys. (J. Phys. D), Ser. 2, *1*, 1651.

Maynard, P. V., M. Elstein and J. A. Chandler (1975), The effect of copper on the distribution of elements in human spermatozoa, J. Reprod. Fert. *43*, 41.

Mehard, C. M. and B. E. Volcani (1975), Evaluation of silicon and germanium retention in rat tissues and diatoms during cell and organelle preparation for electron probe microanalysis, J. Histochem. Cytochem. *23*, 348.

Millonig, G. (1966), Model experiments on fixation and dehydration, in: Proc. 6th Int. Congr. Electron Microscopy, Kyoto, *2*, 21.

Moor, H. (1973), Cryotechnology for the structural analysis of biological material, in: Freeze etching techniques and applications, E. L. Beneditti and P. Favard, eds (Société Française de Microscopie Électronique, Paris), p. 11.

Moreton, R. B., P. Echlin, B. L. Gupta, T. A. Hall and T. Weiss-Fogh (1974), Preparation of frozen hydrated tissue sections for X-ray microanalysis in the scanning electron microscope, Nature *247*, 113.

Morgan, A. J., T. W. Davies and D. A. Erasmus (1975), Changes in the concentration and distribution of elements during electron microscope preparative procedures, Micron *6*, 11.

Nasir, M. J. (1972), Quantitative analysis of thin films in EMMA-4 using block standards, Proc. 5th Eur. Conf. Electron Microscopy, Manchester, p. 142.

Pearse, A. G. E. (1968), Histochemistry, Vol. 1, 3rd edn., (Little, Brown & Co., Boston).

Pearse, A. G. E. (1972), Histochemistry, Vol. 2, 3rd edn., (Little, Brown & Co., Boston).

Pease, D. C. (1966a), The preservation of unfixed cytological detail by dehydration with inert agents, J. Ultrastruct. Res. *14*, 356.

Pease, D. C. (1966b), Anhydrous ultrathin sectioning and staining for electron microscopy, J. Ultrastruct. Res. *14*, 379.

Philibert, J., J. Rivory, D. Bryckaert and R. Tixier (1970), Electron probe microanalysis on electron microscope thin foils using thin standards, Met. Phys. *479*, 68.

Podolski, R. J., T. Hall and S. L. Hatchett (1970), Identification of oxalate precipitates in striated muscle fibres, J. Cell Biol. *44*, 699.

Polliack, A., N. Lampen and E. de Harven (1973), Comparison of air drying and critical point drying procedures for the study of human blood cells by scanning electron microscopy, Proc. of the Workshop on Scanning Electron Microscopy in Pathology, IIT Research Institute, Chicago, Ill., p. 529.

Reid, N. (1974), Ultramicrotomy, in: Practical methods in electron microscopy, A. M. Glauert, ed. (North-Holland, Amsterdam).

Robison, W. L., L. Van Middlesworth and D. David (1971), Calcium, iodine and phosphorus distribution in human thyroid glands by electron probe microanalysis, J. clin. Endocr. Metab. *32*, 786.

Roth, L. J. and W. E. Stumpf, eds (1969), Autoradiography of diffusable substances (Academic Press, London).

Rowse, J. B., W. P. Jepson, A. T. Bailey, N. A. Climpson and P. M. Soper (1974), Composite elemental standards for quantitative electron microscope microprobe analysis, J. Phys. E: Scientific Instruments *7*, 512.

Saubermann, A. J. and P. Echlin (1975), The preparation, examination and analysis of frozen hydrated tissue sections by scanning transmission electron microscopy and X-ray microanalysis, J. Microscopy *105*, 155.

Scherrer, R. and P. Gerhardt (1972), Location of calcium within bacillus spores by electron probe X-ray microanalysis, J. Bact. *112*, 559.

Sjoström, M. and L. E. Thornell (1975), Preparing sections of skeletal muscle for transmission electron analytical microscopy (TEAM) of diffusible elements, J. Microscopy *103*, 101.

Skaer, R. J., P. D. Peters and J. P. Emmines (1974), The localisation of calcium and phosphorus in human platelets, J. Cell Sci. *15*, 679.

Spurr, A. (1969), A low viscosity epoxy resin embedding medium for electron microscopy, J. Ultrastruct. Res. *26*, 31.

Spurr, A. (1972), Freeze-substitution additives for sodium and calcium retention in cells studied by X-ray analytical electron microscopy, Bot. Gaz. *133*, 263.

Spurr, A. R. (1973), Freeze substitution systems in the retention of elements in tissues studied by X-ray analytical electron microscopy, in: Thin section microanalysis, J. C. Russ and B. J. Panessa, eds (Edax Ltd., USA).

Spurr, A. R. (1974), Macrocyclic polyether complexes with alkali elements in epoxy resin as standards for X-ray analysis of biological tissues, in: Microprobe analysis as applied to cells and tissues, T. Hall, P. Echlin and R. Kaufmann, eds (Academic Press, London).

Spurr, A. R. (1975), Choice and preparation of standards for X-ray microanalysis of biological materials with special reference to macrocyclic polyether complexes, in: Biological microanalysis, P. Echlin and P. Galle, eds (Soc. Française de Microscopie Electronique, Paris), p. 287.

Stelzer, R., A. Lauchli and D. Kramer (1975), Interzellulare Transportwege des Chlorids in Wurzeln intakter Gerstepflanzen, Cytobiologie *10*, 449.

Stumpf, W. E. and L. J. Roth (1965), Frozen sectioning below −60 °C with a refrigerated microtome, Cryobiology *1*, 227.

Sumi, S. M. and P. D. Swanson (1971), Limitations of the pyroantimonate technique for localisation of sodium in isolated cerebral tissues, J. Histochem. Cytochem. *19*, 605.

Sweatman, T. R. and J. V. P. Long (1969), Quantitative electron probe microanalysis of rock forming materials, J. Petrol. *10*, 332.

Tandler, C. J., C. M. Libanati and C. A. Sanchis (1970), The intracellular localisation of inorganic cations with potassium pyroantimonate, J. Cell. Biol. *45*, 355.

Tandler, C. J. and A. J. Solari (1969), Nucleolar orthophosphate ions, J. Cell. Biol. *41*, 91.

Thomas, R. S. (1974), Use of chemically reactive gaseous plasmas in preparation of specimens for microscopy, In: Techniques and applications of plasma chemistry, J. R. Hollahan and A. T. Bell, eds (Wiley Interscience, NY).

Thomas, R. S. and J. R. Hollahan (1974), Use of chemically reactive gas plasmas in preparing specimens for scanning electron microscopy and electron probe microanalysis. Proc. 7th Ann. SEM Symp, O. Johari, ed. IITRI, Chicago, Ill. p. 84.

Thornton, P. R. (1968), Scanning electron microscopy (Chapman & Hall, London).

Torack, R. A. and M. Lavalle (1970), The specificity of the pyroantimonate technique to demonstrate sodium, J. Histochem. Cytochem. *18*, 635.

Van Harreveld, A. and F. I. Khattab (1969), Changes in extracellular space of the mouse cerebral cortex during hydroxyadipaldehyde fixation and osmium tetroxide post fixation, J. Cell. Sci. *4*, 437.

Van Lannep, E. W. and H. Komnick (1971), Histochemical demonstration of sodium and chloride in the frog epidermis, Cytobiologie *3*, 137.

Van Steveninck, R. F. M., M. E. Van Steveninck, T. A. Hall and P. D. Peters (1974a), X-ray microanalysis and distribution of halides in *Nitella translucens*, in: Proc. 8th Int. Congr. Electron Microscopy, Canberra, p. 602.

Van Steveninck, R. F. M., M. E. Van Steveninck, T. A. Hall and P. D. Peters (1974b), A chlorine-free embedding medium for use in X-ray analytical electron microscope localisation of chloride in biological tissues, Histochemistry *38*, 173.

Weavers, B. A. (1971), Combined high resolution electron microscopy and electron probe X-ray microanalysis and its application to medicine and biology, Micron *2*, 390.

Weavers, B. A. (1973), The potentiality of EMMA-4, the analytical electron microscope, in histochemistry: a review, Histochemistry *5*, 173.

Winborn, W. B., C. M. Girard and L. L. Seelig (1972), Ultrastructural localisation of antimonate deposits in the gastric mucosa, Cytobiologie *6*, 131.

Yarom, R. and J. A. Chandler (1974), Electron probe microanalysis of skeletal muscle, J. Histochem. Cytochem. *22*, 147.

Yarom, R., T. A. Hall and P. D. Peters (1975), Calcium in myonuclei: electron microprobe X-ray analysis, Experientia *31*, 154.

Yarom, R., T. A. Hall, H. Stein, G. C. Robin and M. Makin (1973), Identification and localisation of intraarticular colloidal gold: ultrastructure and electron examinations of human biopsies, Virchows Arch. *15*, 11.

Yarom, R. and U. Meiri (1973), Pyroantimonate precipitates in frog skeletal muscle. Changes produced by alterations in composition of bathing fluid, J. Histochem. Cytochem. *21*, 146.

Yarom, R., P. D. Peters, M. Scripps and S. Rogel (1974), Effect of specimen preparation on intracellular myocardial calcium, Histochemistry *38*, 143.

Further reading

(a) *Metallurgical and mineralogical specimen preparation*

Goodhew, P. J. (1972), Specimen preparation in materials science, in: Practical methods in electron microscopy, Vol. 1, A. M. Glauert, ed. (North-Holland, Amsterdam).

Goldstein, J. I. and H. Yakowitz (1975), Practical scanning electron microscopy. Electron and ion microprobe analysis (Plenum Press, New York).

(b) *Biological specimen preparation*

Chandler, J. A. (1975), Electron probe X-ray microanalysis in cytochemistry, in: Techniques of biochemical and biophysical morphology, Vol. 2, D. Glick and R. Rosenbaum, eds (John Wiley and Sons, New York), p. 308.

Echlin, P. and P. Galle, eds (1976), Biological microanalysis (Société Française de Microscopie Electronique, Paris).

Glauert, A. M. (1974), Fixation, dehydration and embedding of biological specimens, in: Practical methods in electron microscopy, Vol. 3, A. M. Glauert, ed. (North-Holland, Amsterdam).

Hall, T., P. Echlin and R. Kaufmann, eds (1974), Microprobe analysis as applied to cells and tissues (Academic Press, London and New York).

Lewis, P. R. and D. P. Knight (1977), Staining methods for sectioned material, in: Practical methods in electron microscopy, Vol. 5, A. M. Glauert, ed. (North-Holland, Amsterdam).

Reid, N. (1974), Ultramicrotomy, in: Practical methods in electron microscopy, Vol. 3, A. M. Glauert, ed. (North-Holland, Amsterdam).

Wisse, E., W. Th. Daems, I. Molenaar and P. Van Duijn, eds (1974), Electron microscopy and cytochemistry (North-Holland, Amsterdam).

Chapter 5

Specimen analysis

5.1 *Specimen-electron interaction*

When a high energy electron (up to 120 keV in this context) penetrates into a solid, most of the energy is initially lost through ionisation of the atoms of the specimen. The phenomena that result from this have already been described in Chapter 1 and include reflection of primary electrons, secondary electron emission, X-ray excitation, beam induced conductivity and heating of the specimen (Fig. 1.1). Thornton (1968) discusses the various processes accompanying electron penetration in detail. The phenomena which particularly concern us here are electrostatic charging, specimen heating, contamination, primary and secondary electron scatter and damage to both the matrix (main specimen body) and the elements being analysed.

5.1.1 *Electrostatic charging*

Whereas most mineral and metal specimens are electrically conducting, most biological materials are non-conducting, and techniques have already been referred to for mounting specimens and coating them with suitable conducting materials such as carbon or aluminium (§ 4.1.3). With thick specimens, where transmission imaging is not required and the image resolution is not very great, aluminium may be used, evaporated *in vacuo* from a tungsten filament, providing that aluminium itself is not being analysed. With ultra-thin specimens for TEM analysis the image resolution is decreased too much by a coating of aluminium and about 10 nm of carbon is preferred, again evaporated *in vacuo* from pointed carbon rods. It is even better to analyse sections in the TEM or EMMA with no protective coating at all. Before analysis the electron beam may be spread out over the specimen for a minute or so to form a very thin (~ 1 nm) layer of carbon contamination that helps stabilise the section. Some analyses have been successfully performed on sections mounted on metal grids with neither supporting film nor conducting coat. The electron beam is very slowly and carefully focused to the small region of interest in the section and the beam current slowly increased to the required level. The value of this procedure depends on the type of specimen, however, as well as on the operating conditions, and conducting films are normally required to prevent electrostatic charging.

An interesting phenomenon observed with thin sections is that when the electron beam is defocused, so that part of the beam is striking the surrounding metal grid bars, no charging occurs. When the beam is focused onto the section and away from grid bars charging is then sometimes observed. Presumably, electrons are scattered from the grid bars when the beam is defocused, neutralising the charge on the specimen.

The importance of thoroughly coating specimens all over, especially thick specimens, is stressed. Small uncoated areas can give rise to odd local charging effects which can affect both the electron imaging and the analysis. For thick specimens the conducting coat must have a suitable path to earth, often achieved by painting a line onto the specimen holder with colloidal graphite or metal paste (§ 4.1.3).

A serious consequence of electrostatic charging has been observed by Chandler (1973) in a thin specimen of glass containing sodium and potassium. These ions diffused in and out of the electron beam as if the specimen behaved like a sponge in which ions were free to drift. Borom and Hanneman (1966) showed a similar effect in thick specimens of glass and Hodson and Marshall (1970) have observed the loss of sodium and potassium during electron irradiation of a thin membrane containing salts of these elements. Thus care has to be exercised when focusing the electron beam onto small regions containing free ions, particularly when examining frozen sections containing diffusible electrolytes. This phenomenon places an upper limit on the electron beam current density that can be used with certain specimens.

5.1.2 *Specimen damage*

Electrons which interact elastically with the specimen, and so lose no energy, produce no damage, while inelastic electron interactions can produce permanent changes. These changes may be *irradiation damage*, which is electron dose dependent, or *thermal damage*, in which the specimen heating is affected by specimen mass and conductivity. The degradation of the specimen arises from bond rupture and may be observed in some crystalline materials by the loss of reflections in a diffraction pattern (Grubb and Keller 1972; Dobb 1972).

The amount of energy which is given up by electrons to the specimen in the form of heat depends on the nature of the specimen. Conductive coatings and conducting specimen holders help reduce heating effects, but the most important factor seems to be the conductivity of the specimen itself. In a thick specimen each local area, except at the surface, is in intimate contact with the whole matrix and has effectively a 360° solid angle of emissivity. In a thin section each part of the specimen is very close to the surface and hence to the conducting support or coating film. Occasions do arise, however, when part of the specimen being irradiated is of much greater density than its surrounding matrix and has a high absorbance for electrons. In thin sections this may result in a very rapid temperature rise of the local region, possibly resulting in elemental loss or mass loss (Reimer 1965; Kritzinger and Ronander 1974). Saubermann and Echlin (1975) have found, however, that when using a cold stage the temperature rise of a specimen may be very small during electron irradiation of biological specimens.

The consequences of electron beam irradiation of organic material have been studied by many investigators, (e.g. Bahr et al. 1965; Reimer 1965; Stenn and Bahr 1970; Thach and Thach 1971; Glaeser et al. 1971). Bahr et al. (1965) measured the mass losses that occur in supporting materials and organic embedding media under certain irradiation conditions. They found that irradiation changes in organic materials at conventional voltages of electron microscopy are very rapid initially, levelling off with time. They also observed that halogens, particularly fluorine, are very sensitive to irradiation; that changes can occur in the binding energies of molecular compounds during irradiation; and that very large mass losses can occur in some materials.

Reimer (1965) has discussed the various effects of electron irradiation on organic and inorganic objects and includes a useful diagram describing the various process that may occur. He considers the mechanism of radiation damage in organic matter according to radiochemical theory and discusses the effects of beam current intensity and the type of tissue on bond ruptures, cross-linking effects and loss of H, O and N atoms. As far as microanalysis is concerned two major effects must be considered: the loss of specimen matrix, giving rise to changes in mass thickness, and the loss of elements as they are being analysed. To minimise both of these, conducting coatings are applied and specimens are mounted on conducting bases. In addition, attention must be paid to the operating conditions. There is evidence that there is a reduction in loss of organic material at low temperatures (Hall and Gupta 1974), and thus cold stages can be used to maintain specimens near liquid nitrogen temperatures during analysis. Such a stage is easily fitted in the SEM and EPMA where there is adequate room within the specimen chamber for such a device (Saubermann and Echlin 1975). With the TEM and EMMA instruments, however, the problem of maintaining specimens frozen is far more difficult, although some attempts are being made to design such stages.

Höhling et al. (1971) studied the loss of material from 6 μm sections of kidney homogenates and kidney, cut frozen, freeze-dried and mounted on nylon films. The effect of specimen current and Al coating was examined. Large mass losses were encountered, especially with uncoated specimens.

As in the reduction of electrostatic charging, specimen damage can be kept to a minimum by carefully choosing the correct operating conditions (Agar et al. 1974; see also § 5.2). With thin specimens it is well known that more damage occurs at relatively lower voltages (e.g. 30 or 40 kV) than at higher voltages. This is because the scattering cross-section increases as the electron energy decreases. At very high accelerating voltages (e.g. 1 MV) '*knock-on*' damage may occur in certain specimens, but to date little attempt has been made to use such electron beams for X-ray analysis.

Unfortunately with thin specimens the total quantities of elements being analysed are often very small and it is necessary to operate at relatively high electron beam current densities or with long counting times, or both, to obtain statistically significant X-ray information.

Before a series of analyses is performed for a given element in a specimen some preliminary experiments should be made to determine the tolerance of the specimen to the electron beam. A low beam current is used to generate X-rays from a set region of the specimen and the X-ray signal recorded over a typical analysis time (say 100 sec). The electron beam current is then raised to a new value (say doubled) and the X-ray output again recorded over a time period. In this way it may be determined at what beam current density a decay in the X-ray signal occurs during the period of analysis. For subsequent analysis a 'safe' beam current density is then chosen which is lower than the threshold value but high enough to provide adequate X-ray output. Observations on decay of the X-ray signal such as these should be made on specimens as a matter of course before a routine quantitative analysis.

Ion etching of the specimen can occur if the vacuum in the specimen region is poor (Agar et al. 1974). Gas molecules near the specimen are ionised by the beam and these in turn cause a local ion bombardment resulting in a rapid and serious etching of the specimen.

5.1.3 Contamination

Another result of poor vacuum is the polymerisation by the electron beam of hydrocarbon molecules deposited on the specimen from the vapours from oils and greases in the vicinity of the sample. If the vacuum is very poor then heavy and rapid contamination can result. In the TEM, where the contamination affects the image quality and is immediately obvious, contamination rates of 0.1 nm/sec are intolerable. Methods for measuring contamination rates are described by Agar et al. (1974). With imaging systems having less severe requirements, such as the EPMA and SEM, the problem is not so great.

There are two consequences of contamination. Firstly, the contaminating layer absorbs low energy X-rays emerging from the specimen. Secondly, the mass thickness of the specimen is affected (sometimes drastically in ultrathin sections) thus making quantitative analysis difficult and reducing sensitivity by increasing the background. The effect of contamination on mass thickness is discussed further in § 6.4.1.

To reduce contamination, *anti-contaminators (cold fingers)* are often employed in the instrument. These are copper blades surrounding the specimen, which are cooled externally to −196 °C with liquid nitrogen. Any polymerised hydrocarbons formed by the electron beam are trapped when they condense on the relatively cool blades. If possible, contamination should be avoided altogether by maintaining a high and clean vacuum in the specimen region, but this is not always easy since the specimen and photographic plates are sources of contamination. A full account of the formation of contamination and its prevention is given by Agar et al. (1974). Thin plastic X-ray windows are often employed to separate the vacuum of the specimen region from the X-ray detection system (§ 5.2.6d, § 5.2.7b).

Contamination is often found to be worse when lower electron beam current densities are employed. This may be due to the fact that the region of the specimen being examined has a lower temperature than when high beam currents are used. Providing the specimen can tolerate the extra irradiation, higher beam currents may thus help to reduce contamination during analysis. Attention should always be paid to possible sources of contamination such as poor microscope vacuum, grease or oil on internal parts of the microscope, and finger grease on grids, specimen holders or apertures. The microscope should be cleaned regularly according to routines described in the manufacturer's manual. In general, special care should be taken to ensure that parts irradiated by the electron beam are kept clean.

5.1.4 Electron scatter

As well as being absorbed by the specimen to produce X-rays, cause radiation damage and generate heat, electrons are scattered from the surface of the specimen and strike other parts of the instrument. In consequence an X-ray signal from these parts enters the detector (Fig. 5.1). The degree of scattering from the specimen depends on the accelerating voltage, beam current and scattering power of the specimen itself; the denser regions of the specimen usually scatter electrons more effectively than the less dense regions.

It should be realised that, even with a finely focused electron beam, a small fraction of electrons will be irradiating a much large area of the specimen and its support through random scattering in the microscope

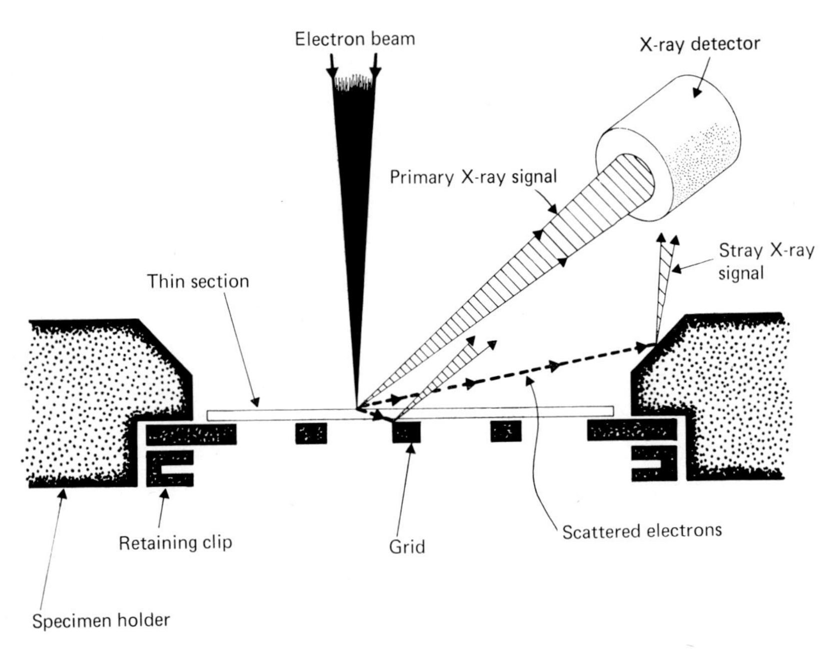

Fig. 5.1. The production of spurious X-rays by electrons scattering onto the specimen support. (Courtesy of John Wiley and Sons, Inc., New York.)

column. These electrons give a small contribution to the spurious X-ray signal. The consequences of electron scatter on quantitative analysis are described in Chapter 6.

5.2 *Choice of operating conditions*

All the factors just discussed in § 5.1 govern the choice of operating conditions for the microscope, but there is no general rule for operation of the instrument for optimum conditions for analysis, since each specimen has different requirements.

5.2.1 *Accelerating voltage*

In the electron microscope the choice of accelerating voltage is governed by the following criteria:

5.2.1a *For transmission imaging (TEM)*

A higher accelerating voltage (say 100 kV) gives better penetration of thick specimens and better image resolution as a result of lower chromatic aberration. A lower accelerating voltage (say 40 kV) gives greater contrast in the image but greater risk of specimen damage – particularly in biological specimens. Thus thick specimens require higher voltages for penetration and specimens lacking contrast require lower voltages. Agar et al. (1974) discuss the choice of accelerating voltage for imaging in the TEM.

5.2.1b *For surface imaging (SEM)*

SEMs are usually operated in the range 1–30 kV and the voltage chosen will depend on the nature of the specimen and the information required. It is not necessary to operate at the same accelerating voltage for both imaging and analysis but this is often the procedure adopted. The choice of accelerating voltage for specimen imaging in the SEM is discussed by Thornton (1968), Oatley (1972), Hearle et al. (1972), and Goldstein and Yakowitz (1975).

5.2.1c *For analysis*

A higher accelerating voltage allows a greater gun brightness and thus higher electron beam currents. For bulk specimens, higher accelerating voltages result in greater penetration of electrons and hence worse spatial resolution (§ 3.1.1).

The ionisation cross-section (relative ionisation probability) is $Q = (1/E_0E_c) \log (E_0/E_c)$, where E_0 is the energy of the primary electron beam and E_c is the critical excitation potential (§ 2.4.1). Q reaches a maximum for most elements between 2.5 and 3 times the absorption edge energy E_c, and to a large extent determines the optimum choice of accelerating voltage for bulk specimen analysis. The efficiency of production of X-rays in a bulk specimen depends, however, on the depth of penetration of electrons and is found to increase with accelerating voltage. Green and Cosslett (1968) have shown that the X-ray production for characteristic radiation is $\propto (E_0 - E_c)^{1.63}$. Thus for the optimum X-ray yield there needs to be a compromise; for maximum ionisation cross-section, a low value of E_0E_c is required, while for high yield of X-rays, $E_0 - E_c$ needs to be large. In practice a value of $E_0/E_c > 2.7$ is often chosen.

For thin specimens penetration problems are less important and, since higher gun brightnesses are available at higher accelerating voltages, values of E_0 far in excess of 3 times E_c are often used. An added advantage is the improved X-ray spatial resolution with higher accelerating voltages in thin specimens, the opposite of that found in bulk specimens (§ 3.1.1).

The generation of white radiation (background) in the specimen also depends on accelerating voltage. For thin specimens background is actually reduced in the low energy X-ray range at higher accelerating voltages and hence peak-to-background ratios are improved. Wide-angle electron scatter from the specimen is also reduced at higher voltages, again reducing background intensity.

The choice of accelerating voltage, within the range of 20–100 kV in a TEM, depends on a number of factors and preliminary experiments are necessary for each specimen type to determine the best voltage as follows:

(i) Determine what range of voltages is possible for suitable imaging of the thin specimen (sufficient resolution to allow subsequent localisation for analysis).

(ii) Determine the maximum current density available at each voltage.

(iii) Determine the best probe diameter available at each voltage for the required beam current density.

(iv) Determine the tolerance of the specimen to electron beam damage at each voltage.

(v) Determine the optimum X-ray yield for each element of interest at each voltage (using thin standards).

(vi) Determine the effect of electron scattering on X-ray background at each voltage.

After a brief series of preliminary experiments the best choice of voltage for each analysis and each specimen will become apparent.

The same rules apply for thin specimens viewed in the SEM but the factors governing image resolution, spatial resolution and electron beam damage in thicker specimens are somewhat different. The voltages employed in a typical SEM range from 1–30 kV. Electron beam damage may be considerable in a thick non-conducting sample at these low voltages. The schedule for choosing SEM voltage for a thick (1–10 μm) specimen is usually similar to the TEM schedule for thin specimens given above. Hall (1971) discusses the effects of accelerating voltage on thick biological specimens in detail.

5.2.2 Beam current

Gun brightness increases with accelerating voltage and the amount of beam current (i) that can be focused onto a probe diameter, d, is given by the relationship: $i = k\,d^{8/3}$ where k is a constant proportional to the accelerating voltage. Thus for analyses requiring the detection of trace elements in small regions of the specimen the limit of detectability may depend on the current available in the given probe diameter.

The choice of beam current is made in a preliminary series of experiments as follows:

(i) Determine the maximum beam current density available at each accelerating voltage.

(ii) Determine the smallest probe diameter possible with each beam current (at each accelerating voltage) and decide what current density is required for the analysis.

(iii) Determine the tolerance of the specimen to electron beam irradiation by performing X-ray decay measurements as described in § 5.1.2.

(iv) Determine the X-ray scattering background at each beam current density.

These measurements will then indicate what beam current density at a particular accelerating voltage is most suitable for any particular specimen.

The beam current is measured in some instruments with a Faraday cage which can be moved into the path of the electron beam. Alternatively fluorescent screens that detect beam current for exposure purposes may be calibrated to read actual beam current. If neither a Faraday cage nor a

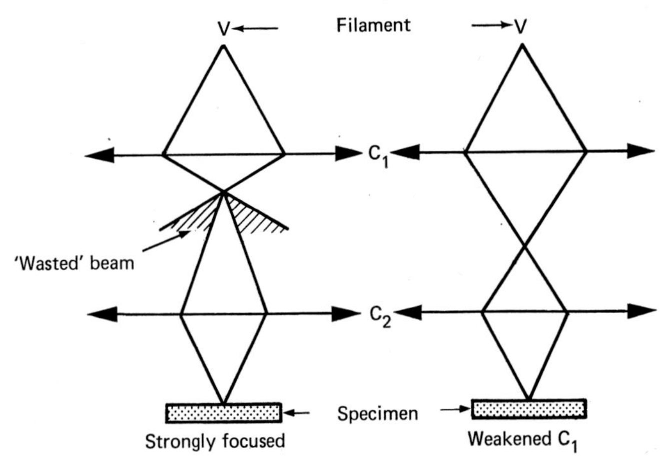

Fig. 5.2. The effect of increasing beam current at the specimen by weakening the first condenser lens. Previously excluded electrons are now included in the final probe, but the probe diameter may be increased.

fluorescent screen detector is available, a photocell may be used to detect screen brightness and this will give an indication of total electron beam current.

Typical current ranges used for analysis are between 10^{-10} A and 10^{-6} A. The beam current may be varied in 3 ways (at a given accelerating voltage),
(i) by changing the condenser aperture (see § 5.2.3 below),
(ii) by changing the strength of C_1 (first condenser lens).
(iii) by varying the gun bias.

Weakening C_1 allows more of the illumination from the gun to be brought to the specimen by reducing the divergence but causes the probe diameter to enlarge (Fig. 5.2). Weakening the gun bias allows a greater source brightness but weakens gun saturation and possibly causes shorter filament life. All three adjustments may be made together to achieve optimum brightness at minimum probe diameter.

Pointed filaments (see Appendix) may be used to achieve improved brightness at the same probe diameter, or reduced diameter at the same brightness (Bowman and Hardie 1972). In the SEM and STEM, field emission sources are also available (Agar et al. 1974).

5.2.3 Apertures

The electron beam diameter is dependent on the apertures employed in the condenser system as well as on the gun current (§ 5.2.2). Apertures may be inserted immediately in front of the gun, just below the anode plate, or in the normal condenser position just below the second condenser lens. The size of aperture affects the final total beam current and the diameter of the probe at the specimen. A 400 μm aperture, for example, situated in the condenser 2 position in the TEM allows a high total beam current to reach the specimen but only with a large probe diameter. A 50 μm aperture, however, greatly reduces the beam current and the beam diameter. A range of apertures should be made available (e.g. 50, 250 and 400 μm).

The objective aperture is usually withdrawn during an analysis in the TEM or EMMA and the anti-contaminator blades may be removed temporarily from the microscope to help reduce background X-rays due to scattered electrons striking these parts. If low levels of contamination are of primary importance, however, the anti-contaminator blades must be retained.

In the SEM the focusing aperture has a similar range of diameters to the condenser aperture in the TEM.

5.2.4 *Length of analysis*

Specimen beam damage (§ 5.1.2) determines the choice of accelerating voltage, of beam current and of the time of the analysis. Preliminary experiments on the specimen should be performed as follows:

(i) Determine the effect of time of analysis on the decay of the X-ray signal (§ 5.1.2) at voltages and beam currents selected as described above.

(ii) Determine the amount of contamination and specimen damage produced in time with the selected voltage and beam current. Since the main difficulty resulting from contamination is the change in specimen mass thickness, the white radiation should be monitored (§ 5.3.4f, § 5.3.5b) over a time period. If the white radiation count rate increases over this counting time period (say 100 sec) with a constant beam current, then contamination may be suspected. If the count rate decreases then erosion of the specimen may have occurred. The rate of contamination can be measured as described in § 6.4.1.

(iii) Determine the time required to produce the necessary X-ray signal.

The accumulated X-ray signal depends on a combination of beam current and time at any given accelerating voltage for a given specimen. Doubling the time of analysis should be equivalent to doubling the beam current. It may be easier to increase the time if the specimen is temperature-sensitive, but irradiation damage may be time-dependent and practical considerations of convenience may govern the maximum time possible for an analysis.

Typical counting times are of the order 10–100 sec, although they may be longer in a static analysis where trace elements are to be detected. In a scanning 2-dimensional analysis the time of analysis may be much longer (in the region of $\frac{1}{2}$ hr) in order to integrate the X-ray display so as to accumulate sufficient X-ray counts in a mapping display (§ 3.4.1c).

5.2.5 Magnification

In every instrument the magnification of the image has no effect on the analysis since all the magnification takes place after the specimen has been irradiated. However, for convenience, a magnification is chosen to assist in the analysis.

In the TEM and EMMA instruments the transmitted electron probe is visible on the screen during the analysis. This should be magnified with the magnification control to approximately half fill the screen. This allows the region of analysis to be observed during the analysis, so as to check the stability of the probe on the specimen and to observe any specimen damage or contamination which may occur. The procedure also prevents a very small intense beam from burning a hole in the fluorescent screen. At any time the probe may be defocused to allow a view of the whole specimen. In the SEM and STEM such imaging is not possible during a static analysis and the choice of magnification is that which is most convenient for switching to the scanning imaging mode. Again the magnification does not affect the analysis.

5.2.6 Crystal spectrometers

5.2.6a Choice of spectral line

Each element produces a range of X-ray lines during electron bombardment (§ 2.3). In general the K lines are most intense, with the L lines, M lines and N lines having progressively less intensity. In the TEM and EMMA instruments the accelerating voltages are high enough to give good yield of most of the K lines (provided $E_0 > 2.7\ E_c$, § 5.2.1c). However, the detector system cannot always detect such lines if they have too great an energy. In practice the governing factor is the diffracting crystal which is only able to reflect a certain wavelength range. The wavelength (or energy) range for a number of crystals was shown in Table 3.1. When a choice of crystals is available then a choice of X-ray lines from a given element may also be available. Fig. 3.19 shows the spectrometer chart used on the EMMA crystal spectrometers. The ZnK_α line can be detected at a wavelength of 0.144 nm with the LiF crystal or the ZnL_α line at 1.23 nm with the mica crystal. In practice the K_α line is far more intense than the L_α line and is chosen for analysis. Whenever the K_α line falls within the range of a diffracting crystal it is generally chosen. The wavelength of the very heavy elements (§ 2.3) is so short that the K_α line does not fall within the diffracting range and so the L_α line or even the M_α line is used. When both K_α and L_α lines can be detected on a range of crystals, such as As, Ge, Ga, Zn, Cu, etc., then both lines may be measured for intensity and peak-to-background ratio, using standards to achieve the best diffracting conditions.

5.2.6b Choice of crystal

It is possible that an element such as Ca can be detected either on a LiF or an ADP crystal, using the K_α line in each case. A standard is used to adjust for the best peak-to-background ratio and X-ray intensity for this line from each crystal. When trace elements are to be detected then low background levels may be more important than actual intensities.

A troublesome effect may occur when attempting to analyse certain elements with chosen crystals. Some of the X-rays emitted from the specimen may cause X-ray fluorescence in the crystal, and this fluorescent radiation may introduce a high background radiation. For example, the K_α lines of Fe, Ni or Cu may cause fluorescence of potassium in a KAP crystal (potassium acid phthalate). This may produce an undesirably high background if the operator is attempting to detect low levels of another element such as Si. It may then be necessary to use a different diffracting crystal such as PET or EDT (see Table 3.1).

5.2.6c Spectrometer pressure

If the analysis is performed with the crystal spectrometers under vacuum (§ 3.2.1c) then the lower energy X-rays from the lighter elements ($Z = 11$–20) will not be attenuated. However, scattered electrons may travel through the vacuum from the specimen to the detector to increase the general background. When heavier elements are being detected it is possible to operate the crystal spectrometers under atmospheric pressure to filter out scattered electrons, since the high energy X-rays will not be attenuated. For the detection of elements in the atomic number range 11–15 a vacuum of at least 0.1 Torr is required.

5.2.6d X-ray window thickness

The thin plastic window separating the spectrometer and microscope vacuum (§ 3.2.1c) may be varied in thickness according to the energy of the X-rays being detected. The attenuation of X-rays of different energies by different materials of varying thickness is illustrated in Fig. 5.3. Thin windows have the advantage of allowing penetration of lower energy X-rays while thicker windows have the same effect as atmospheric pressure and filter out stray electrons. In practice windows of 2 μm thickness are often used for general purpose work with thinner windows where greater sensitivity of detection of lighter elements is required.

Thin windows of polycarbonate are mounted on the appropriate support with an adhesive. For example, 2 μm windows of Makrofol ($C_{16}\,H_{14}\,O_3$) (see Appendix) are mounted on aluminium conical supports for the EMMA instrument in the following way. An adhesive consisting of a mixture of Bostik (see Appendix) with acetone or amyl acetate, in the ratio 1:1, is painted around the end of the cone (Fig. 5.4a). Three-millimeter diameter discs of Makrofol sheet are then punched out with an ordinary paper punch or cork borer, and are laid flat on the bench with the aluminised side downward. The end of the cone is pressed lightly down to make contact with a disc (Fig. 5.4b). The disc is smoothed over the edge of the cone (Fig. 5.4c) and excess adhesive is wiped carefully off. The adhesive is allowed to dry at room temperature for 4 hr or in an oven at 40 °C. Care should be taken to avoid leaving finger grease on the cone surface in the region where it fits into the microscope column. The cone may be wiped clean with acetone. An alternative position for the window in this cone is shown in Fig. 5.4d. Here the window has a greater diameter but it is further from the specimen and is less likely to be irradiated by the scattered electrons which produce X-ray background.

X-ray windows of other materials are made in a similar fashion. The life of a window depends on the pressure difference across it and on the degree of scattered electron irradiation it has suffered, since irradiation causes it to become brittle.

Extremely thin (100 nm) X-ray windows are supported on grid meshes

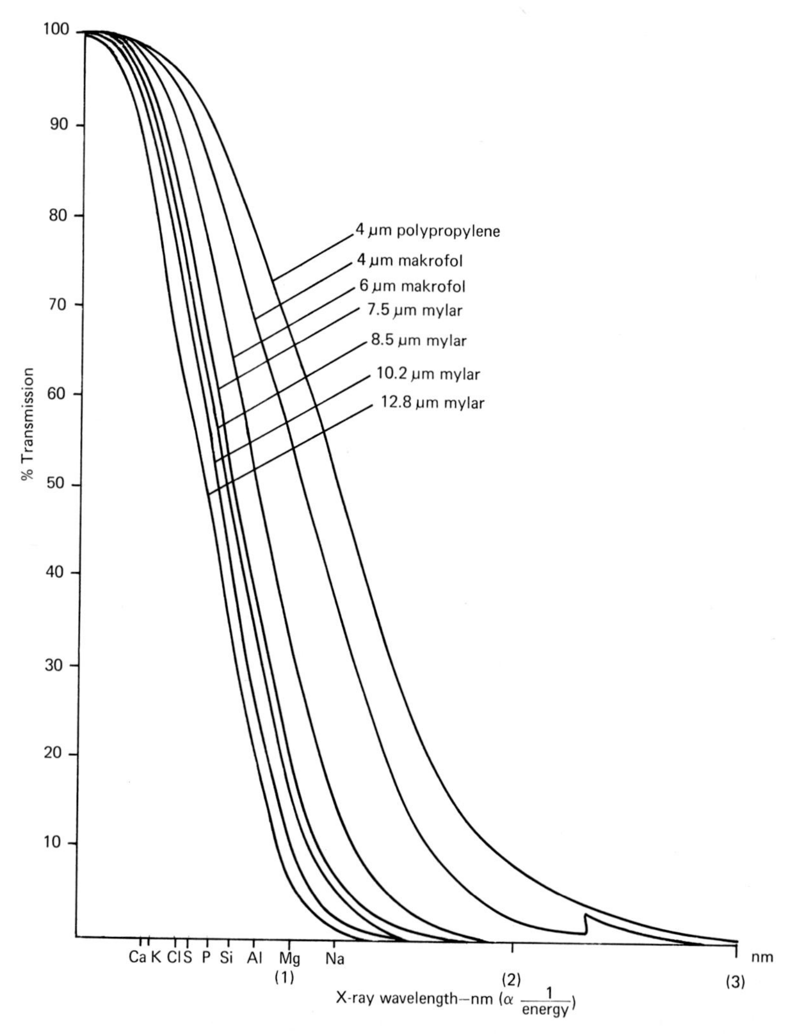

Fig. 5.3. The transmission of characteristic X-rays from different elements through varying thicknesses of window material.

when a pressure difference is to be maintained. Such thin windows allow good transmission of low energy X-rays but are very fragile. When a careful balance of pressure is to be maintained a reasonably large film may be constructed as shown in Fig. 5.4. A suitable film of nylon can be made as described in §4.1.2. A droplet of 0.1 ml of nylon stock solution is pipetted onto a clean water surface. The nylon film is picked up on a suitable frame to be

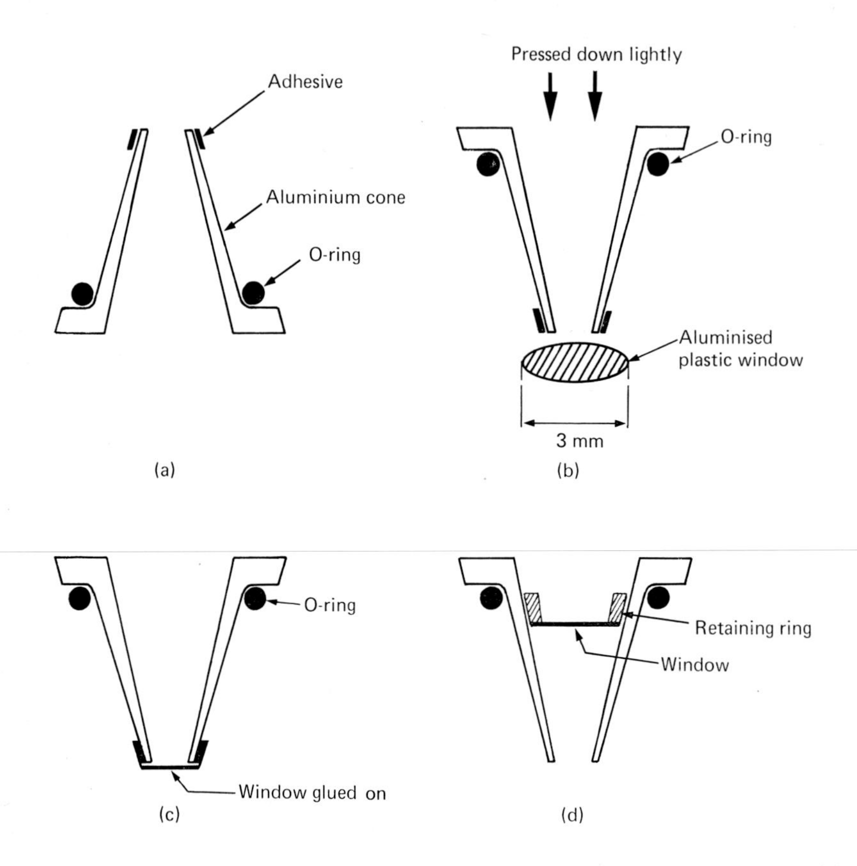

Fig. 5.4. Methods of mounting X-ray windows: (a) the support is thinly coated with adhesive, (b) the support is pressed gently onto thin window material, (c) the adhesive is allowed to harden, (d) alternative position for mounting the window (to reduce electron bombardment).

placed on the end of the cone. Other thin window materials are collodion in acetone and formvar in chloroform (Bradley 1965).

The window in front of the gas-flow proportional counter of the crystal spectrometer (§ 3.2.1b) must be thick enough to withstand the pressure of the gas flow. Windows of 2 μm are generally fitted, although thinner windows can be used if suitably supported.

Thus the choice of window thickness depends on the energy (or wavelength) of the X-rays being detected, and reference to Fig. 5.3 should give some idea of the thickness required. NB: It is very important when replacing X-ray windows to ensure that the thickness of the window material has not changed and that no dirt or excess adhesive is in the path of the X-rays. Such changes will drastically alter the collected X-ray signal and will make subsequent analyses different to those made before changing the window.

5.2.6e Choice of detector collimator

In front of the gas-flow proportional counter window used in a crystal spectrometer there is a collimator which is usually in the form of a thin slit cut in a brass or copper strip. The width of the slit determines the degree of collimation of the focused X-ray beam entering the detector. If the slit is too wide, or if no collimator is present, then some 'off-focus' X-rays will be detected that have not been accurately diffracted, and thus have a wavelength different to the X-rays in the characteristic line. If the slit is too narrow then the focusing arrangement must be very precise indeed to ensure that the diffracted rays enter the detector. Otherwise the intensity of the signal will be reduced. The operator should experiment with different slit widths (usually supplied or easily made) to try and achieve the best peak-to-background count ratio in the spectrometer. Generally slits vary in width from 0.25–2 mm and a commonly chosen size is 1 mm.

The collimator slit should be accurately positioned in the detector to be in the line of focus of the X-ray beam. This may be deduced using a straight edge (e.g. a ruler) in front of the detector with copper X-rays being produced from an irradiated Cu standard with the spectrometer operating at air, and with the front door removed. The spectrometer is first moved to the position for detection of the CuK_α radiation by tuning for a maximum signal from the detector. The straight edge is then held in front of the detector collimator and parallel with the slit such that half of the slit width is covered. If the collimator is accurately positioned in front of the detector the X-ray signal should fall by 50%. Slight adjustment of the slit and repetition of the above procedure will soon determine the correct position for the collimator.

NB: Because of the risk of X-ray irradiation, such adjustments should not be made by inexperienced operators and the assistance of the manufacturer's service engineers should be obtained. Extreme care should be exercised to avoid X-rays striking any part of the operator during this test.

In addition to the collimator used to prevent excessive background X-rays entering the detector, small magnets are often placed near to the detector window to deflect scattered electrons that might otherwise pass through the window and add to the spurious signal. Magnets are also positioned near to the collimators used in the specimen region to prevent electrons entering the spectrometer when X-ray windows are employed in this position. It must be ensured that such magnets are not displaced from their set positions since this will both allow an excess electron signal to enter the detector and may well interfere with the electron optics of the microscope in the specimen region.

5.2.6f Nucleonics

A typical arrangement of the nucleonics associated with the crystal spectrometers is shown schematically in Fig. 5.5. The term *nucleonics* simply describes the electronic arrangement for the collection and processing of electrical signals produced by the X-rays from the specimen. A field effect transistor (FET) is generally positioned within the gas-flow proportional counter of the crystal spectrometer. This delivers the signal to an amplifier and then to a pulse height analyser (PHA). The PHA is a kind of voltage window through which the electrical pulses must pass if they are to be recorded by the ratemeter. A diffracted X-ray entering the detector produces a voltage pulse in the amplifier and voltages v_1 and v_2 in the PHA are set to define the voltage range so as to allow this pulse through. Once this window has been set only X-rays producing a particular pulse height will be allowed through to the ratemeter. This prevents extraneous pulses from entering the ratemeter and so provides a good peak-to-background ratio.

To set the PHA:

(i) Set v_2 to maximum and v_1 to minimum (without allowing a high electronic noise level to be registered).

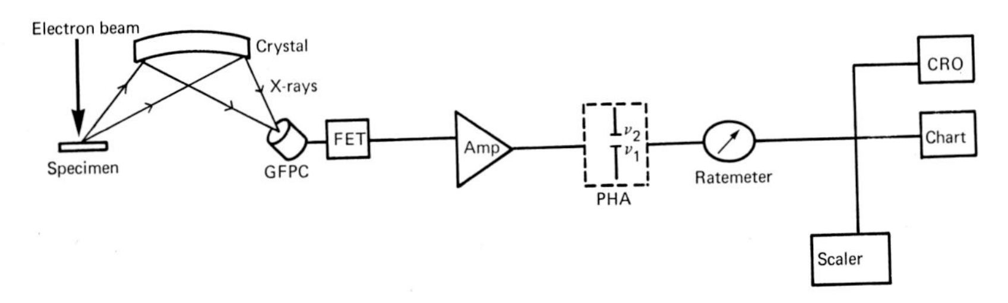

Fig. 5.5. Simple schematic diagram of the nucleonics used for the collection of the X-ray signal in a crystal spectrometer. The spectrometer detector (GFPC) has a field effect transistor (FET) built into it to reduce noise before the signal reaches the amplifier (Amp). Pulse height analysis (PHA) allows pulses of amplitude between v_1 and v_2 to pass through the voltage window into the ratemeter. The final signal is fed to a scaler, to produce X-ray counts in a preset time, or to a CRO or chart recorder.

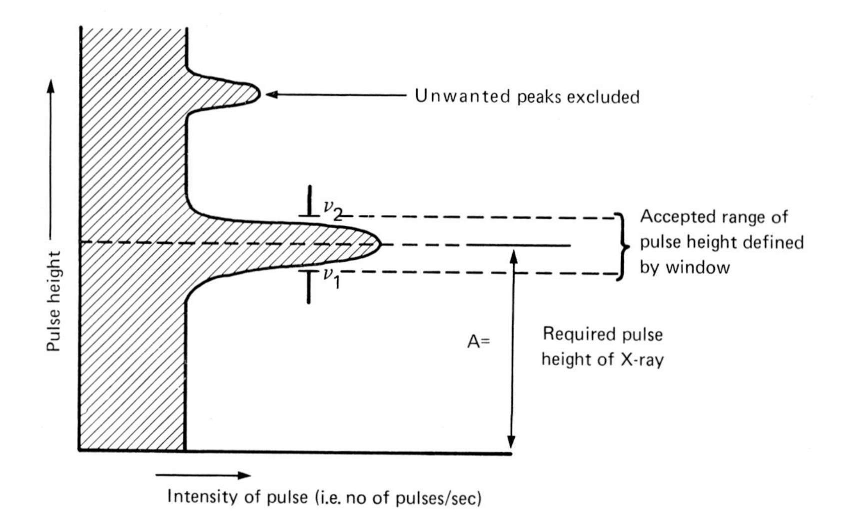

Fig. 5.6. Pulse height analysis using a voltage window. Only pulses of a certain amplitude, A, can pass through the window.

(ii) Tune the spectrometer crystal for a maximum reading of the required X-ray wavelength on the ratemeter.

(iii) Close the window width by lowering v_2 and raising v_1 until the signal on the ratemeter just starts to fall.

The situation is then as shown in Fig. 5.6 where the window defines the accepted height and width of the X-ray pulse. The actual height of the pulse 'A' may be adjusted with the amplifier. Thus the window settings may remain the same for all the X-ray wavelengths to be detected (from many different elements) and the pulse height 'A' adjusted with the amplifier to allow each wavelength in turn to fall inside the window. If too wide a voltage window is chosen there will be too great a background signal included with consequent reduction of sensitivity. If too narrow a window is used the total signal to the ratemeter is reduced. Experience will indicate the correct window width.

The dead-time (§ 3.2.2) of a gas-flow proportional counter of the argon-methane type is about 1 μsec which effectively means that it is possible to detect count rates up to 10^4 counts per second (cps) efficiently. However, one effect of having too high a count rate is that the pulse height is reduced so that if a window setting is chosen for pulses of a particular energy at a count rate of 1000 cps, and then the count rate is raised to 10,000 cps, it will be found that the effective pulse height is reduced ('A' becomes smaller in Fig. 5.6) and the pulses do not pass through the window at the original settings. Thus care should be taken to set the window voltages with a signal from a standard emitting X-rays at the same intensity (count rate) as that expected from the specimen.

The effect of dead-time is to lose recorded intensity, even though X-ray signal intensity is increasing, since the detector is not able to register all the X-rays it receives. Count rates should therefore be kept below 10,000 cps. In practice, with biological samples, count rates are much less than this.

The signal is recorded in the ratemeter (Fig. 5.5) and may be passed to an oscilloscope for ease of detecting low intensities. A chart recorder can plot the signal from the spectrometer if the spectrometer is made to scan through a wavelength range and will thus record peaks of intensity occurring at specific wavelengths. The scaler counts the actual number of X-rays passing to the ratemeter in a given preset time interval. The time of analysis is chosen using the criteria discussed in § 5.2.4.

5.2.7 Solid state detector

Whereas the crystal spectrometer is used to detect single elements having well defined X-ray wavelengths, the solid state detector (SSD) (energy dispersive analyser) is able to detect and display a range of X-ray wavelengths (or energies) simultaneously (§ 3.2.3).

5.2.7a Choice of spectral line

The SSD is not linearly efficient at detecting X-rays over the whole range. The efficiency of collection of K_α X-rays is poor for the lightest and heaviest elements and best for the range $Z = 20–40$ (§ 3.3.2). At the lower X-ray energies (lightest elements) the X-rays are not able to pass through the thin protective beryllium windows. Some detectors are now being manufactured without the protective windows and these may have greatly increased sensitivity for the lighter elements. Detector window thicknesses of 8 μm or 12 μm are commonly used (§ 5.2.7e).

The higher energy X-rays (from the heaviest elements) pass straight through the detector without being stopped. Thus the choice of spectral line depends on the element being detected. There is little choice for elements of atomic number < 30 since only the K_α line falls in the detectable energy range. However, when heavier elements are being detected, it may be

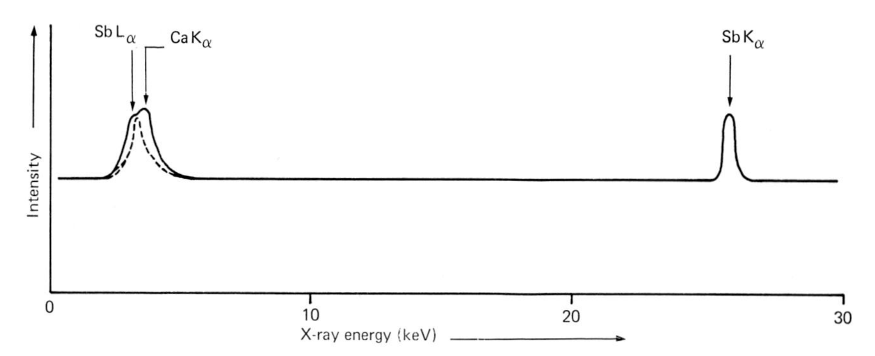

Fig. 5.7. Energy spectrum of calcium and antimony. CaK_α and SbL_α overlap at 3.6 keV, but SbK_α is isolated at 26.35 keV.

necessary to choose the L, or even the M, lines in the spectrum. The following schedule is recommended:

(i) Determine which X-ray lines from the element being detected fall in the range 0–40 keV by analysing a standard of that element at the chosen accelerating voltage.

(ii) Determine which line has the greatest intensity in the spectrum produced by the analyser.

(iii) Determine whether other elements which are likely to be in the sample also produce X-ray lines of the same energy.

(iv) If the most intense line is overlapped by a line from a different element in the sample then select the next most intense line.

An example of (iv) is shown in Fig. 5.7 where Ca and Sb are being detected in the same sample. The most efficiently detected line for Sb is the L_α line, at an energy of 3.61 keV, but the Ca K_α line is very close at 3.69 keV. The second most efficiently detected spectral peak from Sb is the K_α line at 26.35 keV, and here there is no interference from Ca.

The fact that a number of X-ray lines from each element are displayed simultaneously makes this double check method most useful for avoiding ambiguity.

The analyser may be operated to detect and display X-ray energies over varying ranges. Typically these may be 0–10, 0–20 or 0–40 keV. If a small range (say 0–10 keV) is chosen, then more channels (§ 3.2.3) are available in the energy spectrum of the MCA and a better definition of peaks can be made. Intermediate energy ranges such as 10–20 keV or 20–30 keV may also be selected for analysis.

In general, therefore, the choice of X-ray line for detection is that giving the maximum intensity in the energy spectrum and not overlapping with others being emitted from the sample. Detailed X-ray tables are provided by Bearden (1964).

5.2.7b *Window thickness*

The criteria for the selection of window thickness for crystal spectrometers (§ 5.2.6d) apply to SSDs when the detectors are used in conjunction with an electron microscope. The window thickness (in this case that separating the specimen from the detector) determines the attenuation of X-rays passing towards the detector. In many microscopes (especially SEMs), however, no window is used since the SSDs do not generally contain any outgassing materials, such as are found in crystal spectrometers. Thus the SSD may be placed very close to the specimen to increase the solid angle of detection. Again it is very important to ensure that the window thickness is kept constant when a window is changed and that no dirt or excess adhesive attenuates the X-rays. The relative efficiency of the SSD for different elements (§ 6.3) is drastically changed over the energy range if there is some obstruction of the X-ray beam. For example, the presence of some excess adhesive around the edge of the window will tend to attenuate the lower energy X-rays (from lighter elements) and the fall-off in efficiency will be steeper.

The calibration efficiency should be checked with a standard (§ 4.5) such as $AlCl_2$ or $MgSiO_3$ (talc) to determine the slope of the curve in this region after changing a window.

5.2.7c *Pressure at the detector*

The same criteria apply to the SSD as to the crystal spectrometer (§ 5.2.6c) with regard to the choice of air or vacuum between the detector and the specimen. The effect of making the X-rays pass through air before reaching the detector is to suppress the lower half of the energy spectrum as shown in Fig. 5.8. When high energy lines are to be detected atmospheric pressure can help by filtering out scattered electrons and so reducing background.

5.2.7d *Dead-time*

As with the crystal spectrometer (§ 5.2.6f), 'dead-time' is the period when the detector is receiving, or has just received an X-ray and is 'dead' to other incoming X-ray photons. The higher the intensity of X-ray signal (over the

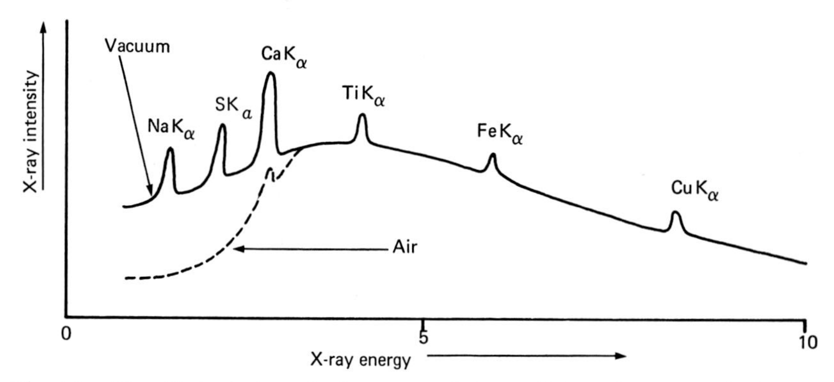

Fig. 5.8. The effect of atmospheric pressure between the specimen and the detector on the low energy region of the X-ray spectrum. Both white radiation and characteristic lines are attenuated before reaching the detector.

whole spectrum) the longer will be the dead-time. If a detector has a dead-time of 10% (i.e. 10% of the total count time lost through the detector being inoperative) and is operated in 'live-time' rather than 'clock-time' as set on the multichannel analyser (MCA), then 10% of the total preset analysis time will be added to the actual time for analysis. Thus an analysis preset to last for 100 sec and having a dead time of 10% will actually count for 110 sec. In general dead-time should not be allowed to reach more than 25% or changes may occur in the energy spectrum (e.g. shift of spectral lines and broadening of peaks). The dead-time, being a function of the X-ray signal intensity entering the detector, is thus determined by the electron beam current irradiating the specimen.

5.2.7e Detector window

As mentioned in § 3.2.3 the SSD is usually provided with a protective beryllium window to seal in the vacuum surrounding the detector crystal. Since X-rays must penetrate this window, it is made as thin as possible. The most common thickness is 12 μm but windows of 8 μm or even 5 μm are also frequently supplied. The advantage of the thinner windows is an increased transmission of the lower energy X-rays, i.e. from 2.5 keV downwards (corresponding to K_α lines from Na–S). Whereas a 12 μm window will transmit only 25% of NaK_α X-rays, a thickness of 8 μm will transmit 60% and a 5 μm thickness ~ 80%. Detectors which are used without windows in front of the crystal collect 100% of the X-rays but such detectors are difficult to operate practically and are not yet in widespread use.

5.3 Performing the analysis

5.3.1 Choice of specimen

Suitable standards (§ 4.5) should first be analysed to tune the detector systems. When a multi-specimen holder is available, standards of particular elements can be placed alongside the specimen. When using crystal spectrometers it is extremely important to have the standard at exactly the same height (within 25 μm) as the specimen to be analysed so that the focusing arrangement of the spectrometers is reproducible (§ 4.5). This is not so critical for the SSD since no focusing of X-rays is involved.

When biological samples are to be analysed it is useful to have the following:

(i) One good ultrathin section for imaging and photography;
(ii) One or two sections suitable for analysis, possibly serial to (i) and stained (perhaps one thin and one thicker);
(iii) One or two sections as in (ii) but unstained (again perhaps two of different thicknesses);
(iv) One standard (of the element to be detected).

It is then possible to analyse specimens for which image resolution is poor. Stained sections allow recognition of areas of low contrast in unstained sections.

5.3.2 Focusing the electron beam

For transmission image analysis of thin specimens in the TEM the image of the region of the specimen to be analysed is brought into the field of view on the fluorescent screen while the microscope is being operated in the conventional transmission mode. The minilens (§ 3.4.4), or focusing lens, is then energised (if it is not already in operation) and the beam is slowly focused onto the chosen region (with the image at a suitable magnification as discussed in § 5.2.5). If charging of the specimen occurs then the specimen should be coated with a layer of carbon as previously described (§ 5.1.1). While this preliminary focusing is being performed the beam current should be reduced using the filament control to avoid excessive heating of the specimen. When a stable focused position is reached the beam current is brought slowly up to the required level.

When using the SEM the transition from a scanned image to a static probe is achieved by a single switching operation. It may be advisable to reduce the beam current in the probe when first performing this operation to minimise specimen heating. The probe diameter can be adjusted and the beam current changed while the static probe is in position on the specimen. If analysis is to be performed with the electron probe scanning the specimen then the diameter and intensity of the probe may likewise be adjusted beforehand.

A movement of the electron probe may be observed if the focusing lens has only recently been energised and is still warming up. It will stabilise after a few seconds. With the TEM or EMMA instruments the image of the specimen region being analysed is clearly visible inside the magnified probe on the fluorescent screen. In the SEM this is not so and for static probe analysis the operator must be sure of the stability of the probe. Fitzgerald (1964) and Reed (1968) discuss probe current stability. The stability of the probe position can be checked when contamination is being produced by observing the contamination mark. A contamination on the specimen extending further than the beam probe diameter indicates that drift has occurred. Most SEMs have crosswires to allow positioning of the probe while viewing the image on the CRO screen and in some TEMs the diffraction beam stop pointer may be used to locate the probe position on the screen. A difficulty with grids and specimen supports made of heavy metals (Cu, Au, Ag, etc.) is that the electron beam may generate very strong X-ray signals when it strikes them. These strong signals may cause the SSD to become inoperative temporarily due to an excessive dead-time in the detector. To avoid this effect a low intensity electron beam is used before the final focused position on the specimen is reached. An advantage of Al or nylon grids is that a far lower background is produced.

Another possible source of error arises if the area of the specimen being analysed is close to a grid bar and the grid bar is between the section and the detector (Fig. 5.9). The signal may be partially or totally prevented from entering the X-ray detector. This will be obvious if the SSD is used to collect X-rays from the whole specimen since no signal at all will be received, but when two spectrometers are being employed simultaneously only one may be receiving a signal (having no obstruction in that direction) while the other is totally or partially shielded by the grid bar. The use of wide mesh grids is recommended to avoid this error. Analyses should not be performed on areas closer than 2 μm to the grid bar, and the grid should always be inserted into the microscope with the specimen uppermost (if the detector is above the specimen) (Fig. 5.9).

The presence of stigmators in the illuminating system of the microscope is

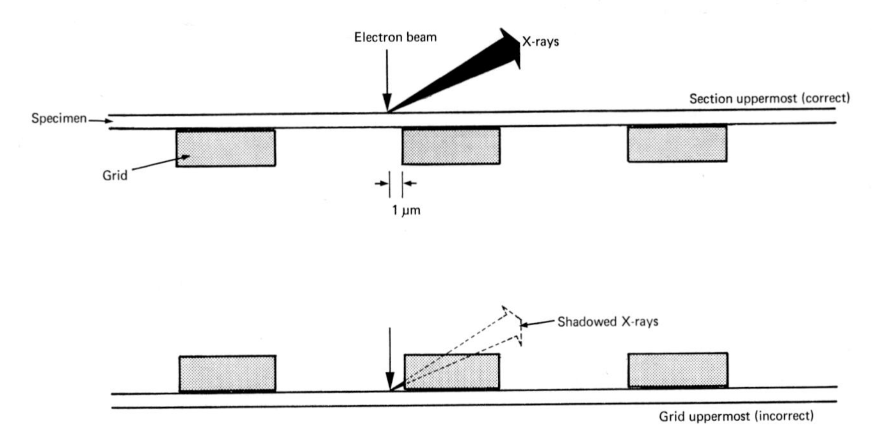

Fig. 5.9. The effect of X-ray shadowing by grid bars. Thin specimens must be placed in the microscope with the section uppermost to avoid attenuation of X-rays.

useful for producing a probe of varying shape. Thus a round probe may be turned into an elliptical shape for focusing onto long objects in the specimen.

In some analyses of ultrathin specimens the region of the specimen to be analysed is smaller than the smallest probe size. For example it may be required to analyse a particle or a subcellular organelle, such as a granule, which has a diameter of 50 nm, while the minimum probe diameter available is 100 nm. Similarly, some cytochemical methods cause the elements of interest to be precipitated in small granules in the section. Alternatively the elemental content of a cell membrane which is only 10 nm wide may be of interest. Only if the electron probe can be placed entirely inside the organelle or particle can valid quantitative comparisons be made. However, provided the qualification is made that the X-rays are emitted from an area equivalent to the beam diameter in each region analysed, then some comparisons may be quoted.

A useful (and sometimes unfortunate) feature of the focused electron probe is its intensity distribution. The typical distribution of intensity across the electron beam focused onto the specimen is shown in Fig. 5.10. It may be seen that a very high percentage of the beam current is in a much smaller region than the actual measured probe diameter. This 'hot spot' can be of value when it is required to analyse something much smaller than the total probe diameter provided the feature is surrounded by a matrix not containing the elements being analysed. If elemental ratios are being determined in discrete particles then the beam 'overlap' will be unimportant if the particles are on a neutral support.

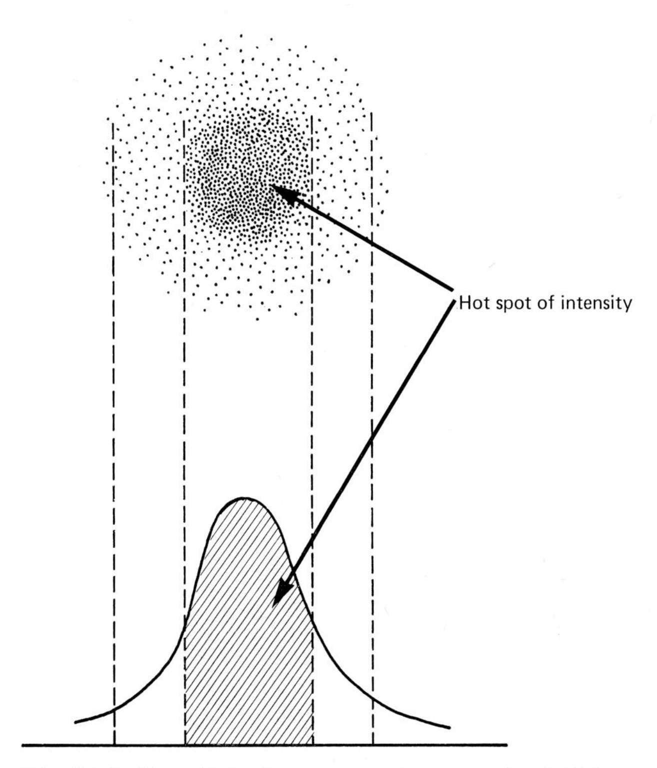

Fig. 5.10. The distribution of intensity across an electron probe. A high proportion of the beam current is contained in the central zone (hot spot) and may be used for analysis of small specimen regions.

If a series of analyses for mass fractions are being performed (§ 6.2.1) then it is essential, when analysing very small areas of the specimen, for the probe intensity profile as shown in Fig. 5.10 to remain constant. If it does not, and specimen regions smaller than the probe diameter (e.g. small particles or precipitates) are being irradiated, then the change in the electron beam density for that small region will give variable results. The focusing arrangements of the microscope should be kept as constant as possible to ensure a constant probe profile, but changes will inevitably occur if, for example, a filament is changed.

To ensure reproducible intensity profiles a standard may be used having particles of constant size and smaller than the beam diameter. A useful standard for this purposes is CaO_2 prepared from a powder as described in § 4.5.2a. The illumination must be adjusted so that such particles give the same X-ray intensity when irradiated with the hot spot both before and after changing the filament for a given set of analysis conditions. The profile of the beam intensity may be measured by moving a single particle of the appropriate size, such as CaO_2, across the probe and plotting the change in intensity of X-ray emission. In some instruments it may be possible to introduce a knife edge into the microscope column to measure the beam diameter (Goldstein and Yakowitz 1975).

5.3.3 Crystal spectrometer

5.3.3a Tuning the crystal with a standard

A standard containing the element to be detected is irradiated with a suitably focused electron beam and with a beam current sufficiently high to produce a count rate of around 1000 cps (approx. 10^{-10} A) using a pure metal standard. If the reading is high enough (say over 100 cps) then the specimen itself may be used as a standard for tuning. Assuming the nucleonics have been adjusted as described in § 5.2.6f, the crystal is brought to approximately the correct angle to diffract the particular X-ray wavelengths required. On some instruments a chart is provided, calibrated in degrees of Bragg diffracting angle (§ 3.4.4) or with a moveable cursor, so that the crystal may be tuned in a similar way to a radio receiver (Fig. 3.19). Other instruments only have an arbitrary numerical scale which is calibrated previously. As the correct angle is reached, a high reading is shown on the ratemeter and/or a bright band of pulses appears on the CRO (Fig. 5.5). The crystal is then set very carefully to the correct angle to obtain the maximum ratemeter reading. Simultaneously, the nucleonics are adjusted as described in § 5.2.6f to achieve the setting for maximum sensitivity. For reproducibility, the crystal should always be advanced to the 'peak' position from the same direction since there may be backlash in the winding mechanism (Fig. 5.11) with the result that the same peak is produced at two different positions. Reproducibility of the setting of the crystal is extremely important in the subsequent analysis because of the reduction in peak height when the crystal is not in the optimum position. For a PET crystal with a resolving power of 7.5 keV set for the K_α line of Si, a movement of 25 μm away from the peak position (on a linear focusing arrangement) can cause a drop of intensity by about 7% and a movement of 50 μm can cause a drop of about 20%. Such errors can occur especially if a spectrometer is used to measure several different X-ray lines in succession. The importance of ensuring that the specimen height is the same for both the standard and the specimen is again stressed

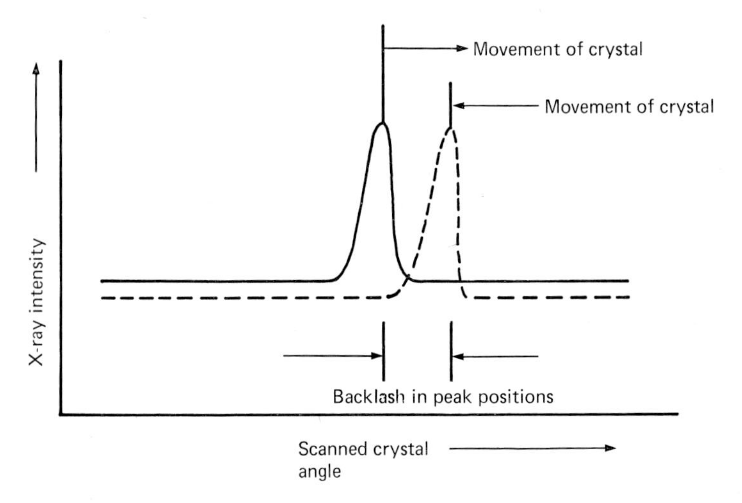

Fig. 5.11. The effect of backlash in the spectrometer drive mechanism. A peak may occur at different settings on the drive scale depending on the direction from which it is approached with the driving gear.

here, since a spectrometer setting will change for a particular wavelength if the specimen height changes.

Servo motors are sometimes employed to preselect wavelength or angle positions in the crystal spectrometer. The angle is set as described above using a standard. The servo motor is then preset so that the crystal may be subsequently returned to the same position automatically by actuating the servo control. The accuracy of reproducibility of such a servo system should be ascertained since inaccuracies in resetting can lead to errors. Servo systems allow a number of angles for a number of elements to be preselected for subsequent rapid selection.

5.3.3b *Tuning the crystal on the specimen*

The electron beam is focused on the chosen region of the specimen to be analysed and then the spectrometer is tuned to the previously determined position. The reading for that X-ray wavelength is then recorded from the ratemeter or from the oscilloscope as a series of pulses. Again, the crystal angle should be approached from the same direction as for the standard to eliminate backlash errors.

Possible sources of error in setting the correct wavelength position (i.e. in setting the correct angle) of the diffracting crystal are:

(i) Backlash in the gear mechanism (see Fig. 5.11).
(ii) Change in specimen height between standard and specimen.
(iii) Non-reproducibility of the servo gear mechanisms.
(iv) Too wide an electron beam diameter. This may produce X-rays from areas 'off focus' with either the Johann or Johansson geometry (§ 3.2.1a). The focusing requirements determine that the probe diameter should usually be less than 50 μm in diameter for highest accuracy.
(v) Large differences in count rate between standard and specimen. The crystal angle should be set with a count rate from the standard in the range 1000–10,000 cps.
(vi) Alignment changes in the spectrometer. If an evacuated spectrometer is vented to air, or vice versa, the casing may slightly distort and cause changes in angle settings for certain wavelengths. Thus the initial alignment must be performed in the same mode as subsequent analyses.

Whenever a series of analyses is to be performed a check should be made on the wavelength settings for the elements concerned. These should not change from day to day but complete stability should not be assumed. Such changes may occur, for example, after renewing a filament when the spectrometers and electron microscope are vented to air.

5.3.3c Measuring the background

A suitable angle is chosen at which to measure the background beneath the peak (Fig. 5.12). A reading at a short distance to the side of the peak may well represent the background but there are a number of possible sources of error.

(i) If the background is sloping (Fig. 5.12b) then an average background is found by measuring the value at either side of the peak and taking the average $(b_1 + b_2)/2$. The background count must be measured for the same period of time as the count at the peak position.
(ii) If another peak occurs close to the measured peak (Fig. 5.12c) this may interfer with the background measurement. A scan across the region by varying the crystal angle will show if such a secondary peak exists.
(iii) The deposition of contamination on the specimen (§ 5.1.3) may change the specimen mass thickness and may change the background reading with time. The rate of contamination must be determined and steps taken to prevent it, if necessary. A method of determining if contamination is likely to occur is described in § 5.2.4.

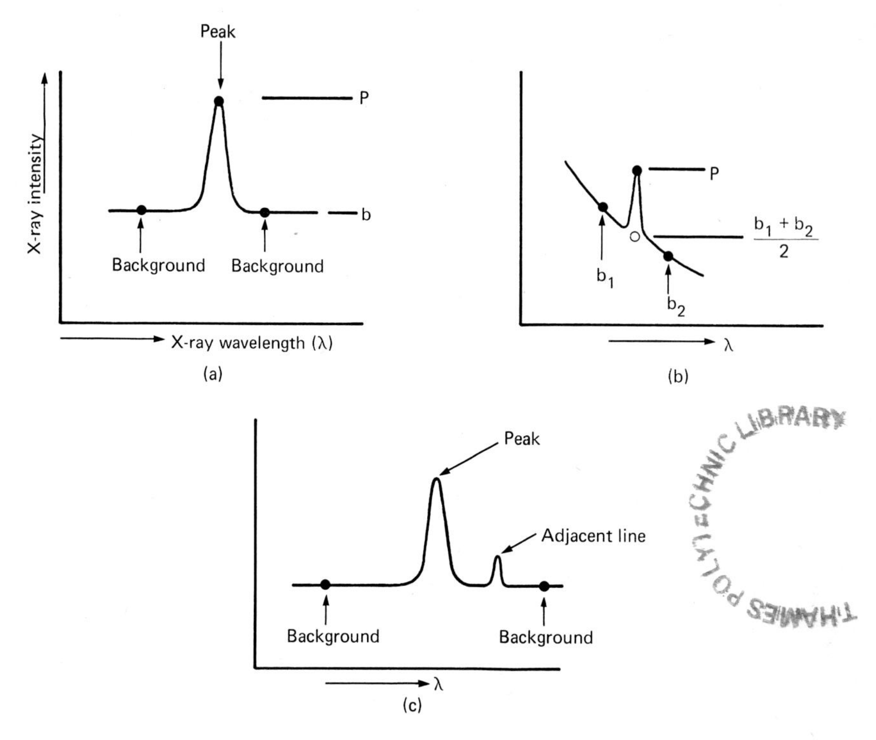

Fig. 5.12. The measurement of background beneath a peak. Readings are taken equidistant on each side of the peak to avoid errors through a sloping background (b). Care should be taken to avoid measuring background at a position near another peak (c).

(iv) Specimen etching in the electron beam may produce a reduction in background (i.e. fall in white radiation) (§ 5.1.2) between counts. Again the possibility of etching occurring must be ascertained beforehand (§ 5.2.4) and the beam current adjusted if necessary.

A method described by Hall and Werba (1969) allows the simultaneous measurement of peak and background counts when two crystal spectrometers are available or when one spectrometer and a gas-flow proportional counter are in use. The amount of resetting of the spectrometers is then reduced. The method is described in § 6.4.1.

By previous measurements it should be determined where the 'true' background occurs, i.e. where the peak is not interfering with the background. In general this is at about 0.02 nm, or 5 times the full width half max. of the peak (Fig. 3.11) (see § 3.3.1), from the peak wavelength position.

5.3.3d Producing an X-ray distribution map in the SEM

The output to the ratemeter during the analysis may be coupled to a CRO operated in synchrony with the scanning raster of the SEM. Thus if the distribution of an element is to be studied in a line across a specimen, areas of high concentration of that element will produce strong signals (if the spectrometer is tuned to detect that element) and consequently spots of high intensity on the CRO. The X-ray distribution on one CRO may thus be compared with an SEM image on an adjacent CRO (§ 3.4.1a).

5.3.3e Analysing the data

The end result of the process of irradiating the specimen with electrons, detecting, collecting and displaying the X-rays as a range of wavelengths, is to produce a set of numbers. The data collected for a particular analysis of one element in a given time period is related to the total number of atoms of that element in the specimen region analysed. It is proportional, over a wide range, to the product iM/a where i/a is the *beam current density* (current per unit area) and M is the total *elemental mass.* Thus for a given set of operating conditions the signal may be considered to be proportional to the elemental mass, or, when the specimen mass thickness is constant, to the elemental concentration. If an analysis is performed at different points on a specimen of constant mass thickness, the X-ray signal provides a relative distribution of elemental concentration.

Since all elements are not equally efficient in X-ray production and the crystal spectrometer is not linearly efficient over its wavelength range, a calibration of X-ray yield for each element is necessary to interpret the numerical data (see Chapter 6).

5.3.4 Energy dispersive analyser

Unlike crystal spectrometers there are no moving parts involved in the energy dispersive analyser (§ 3.2.3) (except when there is a mechanism for moving the detector closer to or further from the specimen). Thus all the operations to be performed are electronic. A number of the most important operations will be described but, since different types of analysers have different controls and are manipulated in different ways, only the basic operations will be described here.

Again, a specimen region is chosen for analysis and the electron beam is focused onto this region. The solid state detector then collects the emitted X-rays. A number of operations may then follow:

5.3.4a *Adjustment of dead-time*

Dead-time is the period when the detector is receiving, or has just received, an X-ray and is 'dead' to other incoming X-ray photons (§ 5.2.6f). The stronger the X-ray signal, the longer will be the dead-time of the detector. Usually, an analysis is performed in the 'live-time' mode to compensate for this loss, but clock-time may be preferred when making certain calibration checks. The dead-time is shown as a percentage on the multichannel analyser (MCA) and should be kept to less than 25% for most work (by reducing the X-ray signal if necessary).

5.3.4b *Setting the energy range*

The MCA is set to collect X-rays over a certain energy range. For example, a setting of 0–10 keV will allow all X-ray energies in that range to be analysed and subsequently displayed.

The K_α lines from atomic numbers 11–32 (Na–Ge) fall in this range. Other energy ranges may be selected, e.g. 0–20, 0–40, 5–10, 10–20 keV, etc. Thus

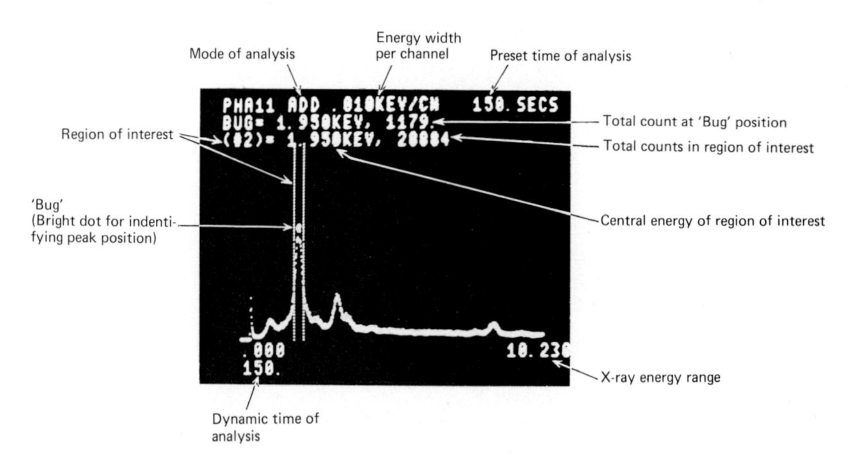

Fig. 5.13. The TV (alpha numeric) display on an MCA. The vertical bars represent an energy band (region of interest) selected for integration, and set here at the fwhm position. The bug is a bright dot, placed here at the top of a phosphorus K_α peak, and indicating the characteristic line energy (1.950 keV) and intensity (1179 counts).

the energies of the X-ray lines to be detected (§ 5.2.7a) determine the energy range to be selected for the analysis (Bearden 1964).

5.3.4c *Performing a conventional analysis*

The X-rays are collected by the analyser and displayed on a CRO, often as an accumulating signal during the time set for the analysis. The final spectrum is examined by identifying each peak and ascertaining its position in the energy range. A typical spectrum recorded from the CRO screen of the MCA by photography with a polaroid or 35 mm camera is shown in Fig. 5.13. The information presented with the spectrum by TV display includes the analysis time; the X-ray band of interest (see below); the peak energy represented by a bright dot (a 'bug') placed at the top of a peak; the peak height, being the number of X-ray counts at the peak position; the running integral, being the total accumulated counts inside the region of interest; and the energy range.

The spectral lines may also be identified by an electronic marker on the screen (Fig. 5.14), which indicates the position, energies and relative intensities of the lines in the spectrum.

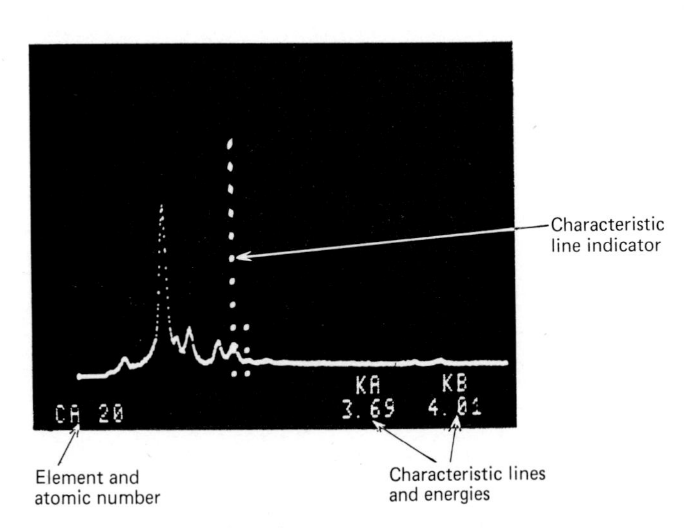

Fig. 5.14. The TV display of an energy spectrum with visual identification of characteristic X-ray lines. The two vertical bars represent the K_α and K_β lines of calcium, the heights of the bars signifying their relative intensity.

5.3.4d *Integrating spectral lines*

The height of a peak in the spectrum allows a quantitative evaluation of that element in the specimen. A more statistically accurate way of measuring peak intensity is to integrate the area under the peak. An energy band (region of interest) is selected on the MCA, as in Fig. 5.13, such that all X-ray counts within the two vertical bars are counted, i.e. the region is 'integrated'. The bars may be set at the full width half maximum (fwhm) position (§ 3.3.1), or at one tenth maximum position, according to the complexity of the spectrum, (one tenth of the maximum position is one tenth of the distance up from the base of the X-ray line to the peak). The integrated value (28884 in Fig. 5.13) is thus obviously greater than the peak intensity value (1179 in Fig. 5.13), giving more X-ray counts and therefore a statistically more accurate result.

Many energy bands may be selected in this way to cover many different peaks. In the same way, regions of interest may be selected to cover background positions to the sides of the peaks. If a computer is attached to the analyser it is often possible to perform an '*integral-minus-background*' process on the spectrum in which each peak is measured, background subtracted, and the final result printed out on a teletype.

Such an output is shown in Fig. 5.15 in which the elements are listed with their respective X-ray line energies, corrected integrated peak values (INT-BG), and background (BG) values. In this experiment the analyser sampled a number of background values at different positions in the spectrum to the

D 446 2673DATA LABEL 50SECS

ID	CENTR	INT-BG	BG 1003
	1001		
NA	1050	73	86
MG	1260	38	104
P	2020	2052	167
S	2320	808	192
CL	2630	519	218
K	3320	446	241
CA	3700	238	234
FE	6405	15	142
CU	8050	15	135
ZN	8650	77	123
	A	B	C

Fig. 5.15. Teletype output for a computerised integration of an energy spectrum for 10 elements. The characteristic line energy (column A) is given together with the X-ray counts in integral-minus-background values (column B) and background values (column C).

sides of the peaks and then the computer extrapolated from a curve running right through the energy range so that the true background was measured below each peak. In some analyses it is possible for the computer to recognise and subtract the background automatically from the spectrum, leaving only the peaks visible on the TV display. In this procedure information is provided to the computer regarding the general composition of the spectrum, the accelerating voltage and other instrumental parameters. The shape of the background is then automatically fitted to the existing spectrum and subtracted.

In addition to being displayed on teletype, the output of information may be stored on magnetic tape, or in the absence of a computer, simply read from the MCA oscilloscope as data values at each spectral position.

5.3.4e Producing an X-ray distribution map in the SEM

When the distribution of one particular element along a line in a specimen being examined in the SEM is required, then an energy band (region of interest) is selected in the MCA to include this one X-ray line, and the total signal within that energy region is collected by a CRO operated in synchrony with the scanning raster of the SEM. In areas of the specimen where the element is present at high concentration large signals are produced from the MCA in the selected energy band, and high intensities are shown on the CRO (see § 3.4.1a).

Instead of the integral of an X-ray energy band, the output signal fed to the CRO for the X-ray image may just be that of the X-ray line intensity measured at its peak. With thin biological sections it is difficult to obtain a sufficiently high X-ray signal to form such an image but some metal or mineral specimens may be examined in this way.

5.3.4f Measuring specimen mass thickness

In many types of analysis (especially biological applications) where the matrix composition is relatively well known, it is possible to measure the specimen mass thickness (see § 2.4.4). This is important in measuring relative or absolute elemental concentration especially in thin specimens where the mass thickness may vary across the specimen (Hall and Werba 1969). The mass thickness of the specimen may be measured by integrating an area of the spectrum (energy band) containing no spectral lines. This represents the white radiation produced by interaction of the electron beam with

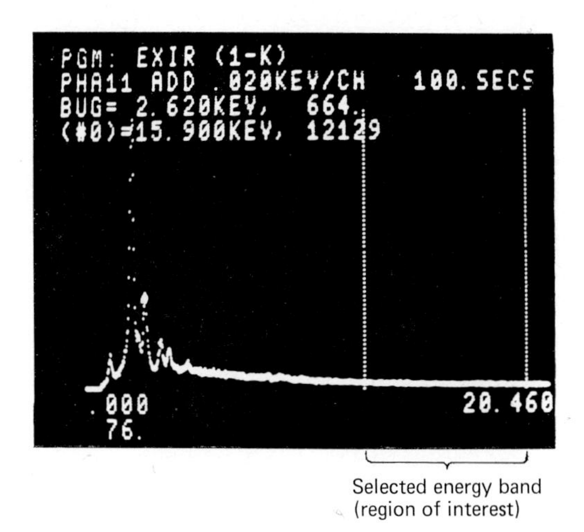

Fig. 5.16. The determination of white radiation from an energy spectrum of a human sperm cell. The energy band for white radiation is set between the vertical bars and includes no spectral lines.

the nuclei of the atoms within the specimen being examined (§ 2.4.4).

For example, consider the spectrum obtained from the head of a human sperm cell shown in Fig. 5.16. The energy band chosen for integration for mass thickness determination is that contained within the vertical bars. The integration of this band is performed in the same way as for any other energy band as described in § 5.3.4d.

To determine the mass thickness of a specimen such as that providing the spectrum in Fig. 5.16 an integration is first performed with the electron beam focused on the specimen, and then with the beam focused on an area of supporting film to the side of the specimen. The difference in these two integrations then represents the true specimen mass thickness (but see also § 6.2.1). The region of the spectrum chosen for integration in determining the specimen mass thickness must not contain any spectral lines. It must also be in an energy range truly representing the white radiation from the specimen. Since, in thicker specimens, absorption of white radiation may occur, an energy band in the upper (higher energy) region of the spectrum is usually selected. In this way the high energy of the white radiation reduces the risk of absorption. On the other hand the energy band chosen must not be so high that the electron energy is unable to generate sufficient continuous quanta within it. A suitable energy band is often in the region of 20 keV when accelerating voltages above 20 kV are used.

5.3.4g Deconvolution of the energy spectrum

When a number of peaks occur in the same energy region such that overlap occurs, it is impossible to tell initially what individual peaks are present (Fig. 5.17). The spectrum is then deconvoluted or '*stripped*' so that each peak may be examined in turn. This can be done in a number of ways, one being by subtracting the peaks electronically by computer according to a 'multiple least squares fit' (often called an ML routine) (Schamber 1973). In this process the computer 'fits' peaks of element 'A', element 'B' and element 'C' in iterative combinations until an exact shape is reached corresponding to that of the complex spectrum. The values of 'A', 'B' and 'C' are then read out automatically by the computer.

An alternative method is to subtract the peak for element 'A' from a complex peak of elements 'A' and 'B' (Fig. 5.18) by analysis of the element 'A' in a 'subtract' mode. Firstly the specimen is analysed to produce the spectrum shown in Fig. 5.18a. It may be noted that the K_α line of Ca and the L_α line of Sb overlap at approximately 3.6 keV, but that the K_α line of Sb is isolated at 26 keV. A pure sample of Sb (or a specimen containing Sb but not Ca) is then irradiated and the analyser set to the 'subtract' mode, i.e.

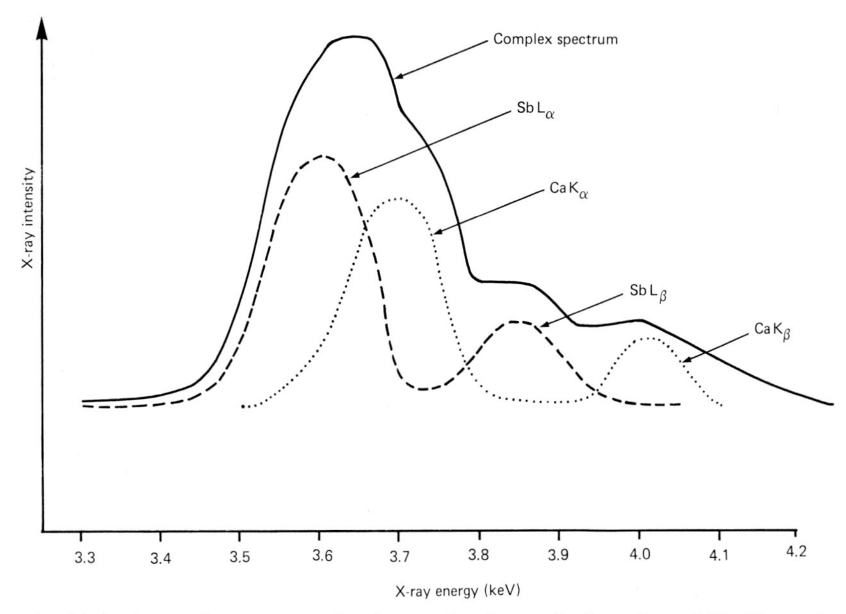

Fig. 5.17. A complex spectrum showing overlapping peaks from Ca and Sb. The peaks can be identified by a process of deconvolution.

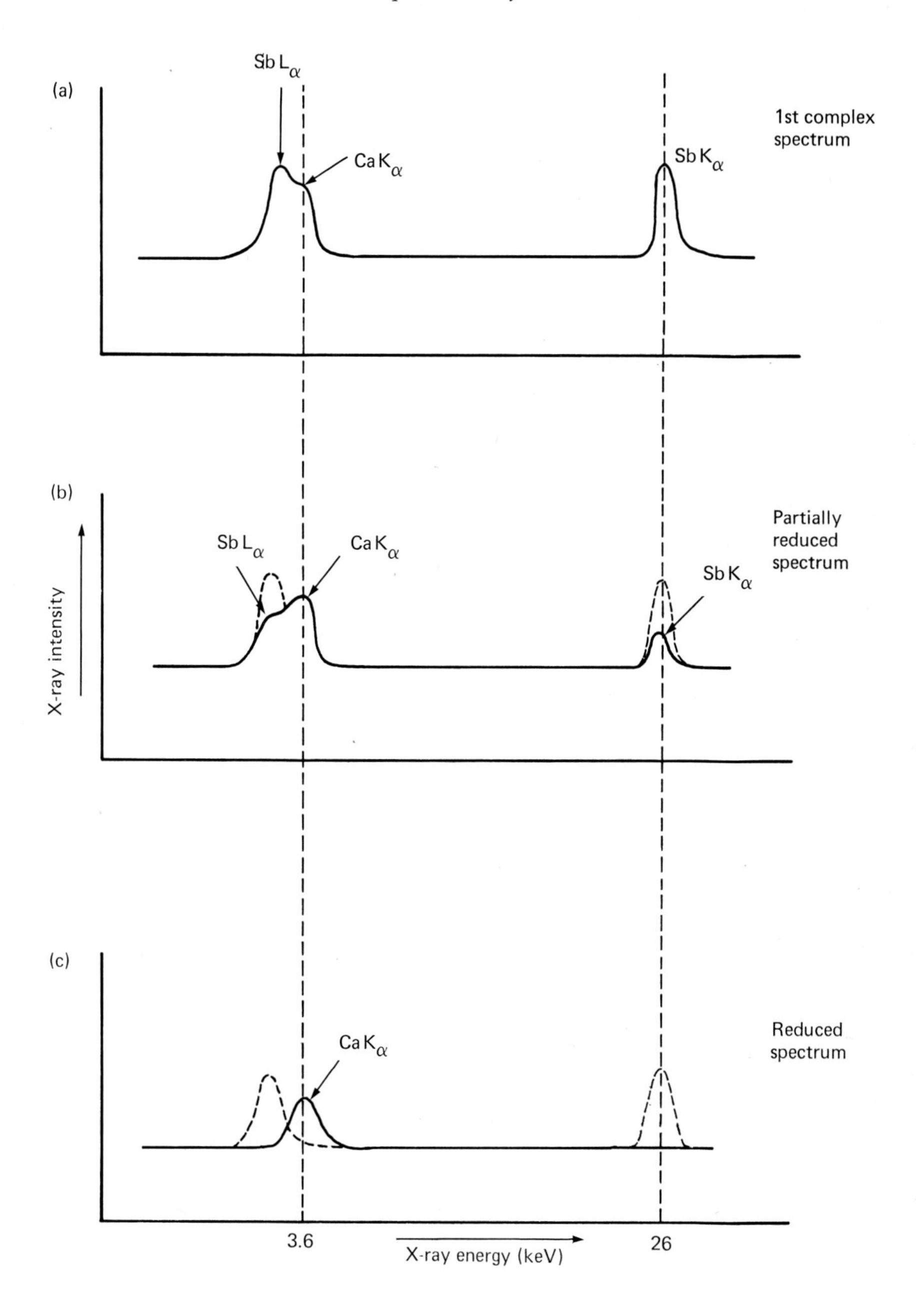

Fig. 5.18. Method of manually subtracting peaks from a complex spectrum (a) of Ca and Sb. A pure Sb standard is used to subtract both the K_α and L_α lines (b) from the initial complex spectrum, leaving behind just the CaK_α line (c).

to subtract the SbK_α and SbL_α lines from the first complex spectrum. The analysis is allowed to continue until the SbK_α line completely disappears, (Fig. 5.18b and c), at which point the SbL_α line will also have disappeared leaving the CaK_α line by itself.

In some analyses the overlapping peaks are automatically separated by synthetically generating a Gaussian peak and applying this peak in a subtractive mode to the original spectrum until the correct background shape remains beneath the position of the spectral peak.

5.3.4h Analysing the data

As with the crystal spectrometer (§ 5.3.3e), the spectrum of X-ray lines is finally processed as a set of numbers such as that shown in Fig. 5.15. These numbers must now be converted into elemental concentrations. For a given element the intensity of X-ray emission is proportional to both electron beam current density and elemental mass over a wide range. Thus the signal (say a peak intensity) is proportional to iM/a where i/a is the beam current density and M is the mass of the element being irradiated. However, the detection efficiency of the SSD is not linear over the energy range (§ 3.3.2, § 6.3) and corrections must be made to the spectral line intensities to provide accurate elemental concentrations. The correction procedure is described more fully in § 6.3. With thicker specimens corrections may also be necessary to account for atomic number, absorption and fluorescence (ZAF) in the specimens. These factors are also discussed in Chapter 6. Such corrections are usually performed by computer methods within the analyser itself.

5.3.4i Calibrating the MCA energy range

The MCA must be adjusted so that the X-ray lines appearing in the energy spectrum are in the correct positions in the energy range on the CRO of the analyser. This entails adjustment of the amplifier shown in Fig. 3.9 and of the 'zero level' in the MCA. The method is as follows:

(i) Place a standard of Al and Cu together in the microscope (an evaporated film of Al on a carbon-coated Cu grid will do).

(ii) Produce an X-ray spectrum on the SSD containing lines of AlK_α (should be at 1.487 keV) and CuK_α (should be at 8.047 keV). This may be done by irradiating both elements in the sample with the electron beam.

(iii) Measure the energy positions of the two peaks as displayed on the

CRO of the multichannel analyser. Suppose that they are (wrongly) at 1.30 keV and 8.50 keV respectively. They are thus separated by an energy difference on the CRO of 8.50–1.30 = 7.20 keV, whereas the correct difference is 8.04–1.47 = 6.57 keV.

(iv) Adjust the *amplifier* until the peaks appear (in a subsequent analysis) with the correct energy difference. They may now, however, still not be positioned correctly in the spectrum.

(v) Adjust the *zero level* until the AlK_α peak appears at 1.47 keV. The CuK_α line should now be at its correct position of 8.047 keV.

(vi) If necessary, further adjust the amplifier and zero level until correct calibration is achieved. This calibration process is not a regular requirement but only when a drift of energy calibration appears to have occurred. It should not be necessary over a long period of operation (months).

5.3.5 Gas-flow proportional counter

The gas-flow proportional counter (GFPC) has an inferior energy resolution to either the crystal spectrometer or the SSD (§ 3.2.2), yet it is useful as a simple means of providing a crude spectrum of X-ray energies and as a means of measuring white radiation for specimen mass thickness determinations.

5.3.5a Producing an energy spectrum

The electron beam is focused onto the specimen to produce a spectrum of X-ray energy emission. This spectrum passes into the GFPC via the thin plastic or beryllium window (generally 2–6 μm). A range of pulses is produced having amplitudes dependent on the X-ray energies. The amplifier (Fig. 5.19) is adjusted to bring the pulse amplitude range onto the CRO screen (Fig. 5.20). The ratemeter records all the pulses falling within the range of the voltage window. This voltage window is set in exactly the same way as described in § 5.2.6f for pulse height analysis with crystal spectrometers. The CRO provides a guide to the nature of the elements present in the specimen, since the pulses having maximum amplitude (i.e. the highest in Fig. 5.20) will be from elements of higher atomic number, and the lowest from the lightest elements. If a chart recorder is available it can be set to record the ratemeter output as the voltage window is slowly moved through the amplitude range (by moving the narrow window vertically in Fig. 5.21).

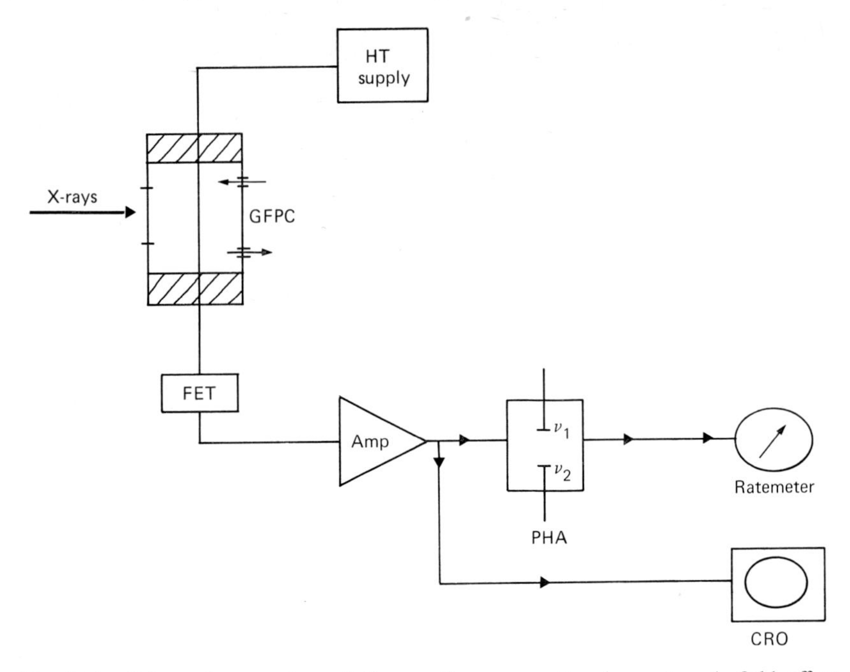

Fig. 5.19. Schematic operation of the gas-flow proportional counter. A field effect transistor (FET) is built into the counter to reduce noise before the signal is amplified. Pulses passing through the voltage window (v_1–v_2) of the pulse height analyser (PHA) are registered on a ratemeter. All pulses from the detector are viewed on a CRO.

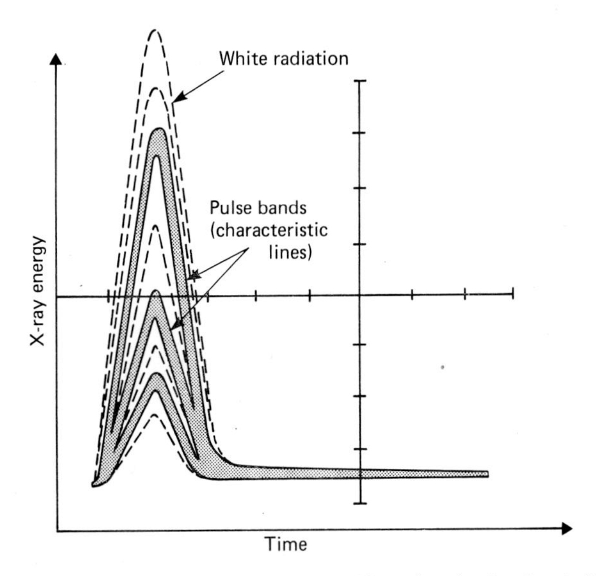

Fig. 5.20. CRO display of pulses from a GFPC. Sharp bands of pulses indicate characteristic lines. Background pulses covering the whole energy range are from X-ray continuum (white radiation).

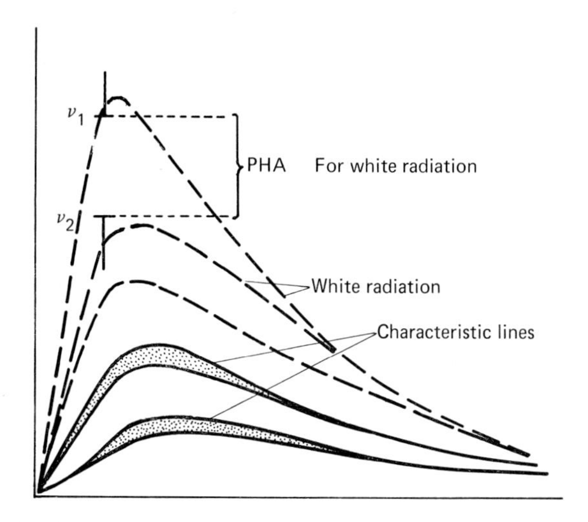

Fig. 5.21. Pulse height analysis applied to the signal from the GFPC. The voltage window, v_1–v_2, allows only certain pulses to reach the ratemeter. The voltage window may be set to collect white radiation (as shown) or individual characteristic lines. v_1–v_2 may be set to any chosen width.

A spectrum similar to that of the SSD will be produced on the chart recorder, having a resolution of the order of 1000 eV.

5.3.5b Measuring specimen mass thickness

In addition to the pulses produced by the GFPC from characteristic X-ray lines, there is also the continuous, or white radiation produced from the specimen during electron irradiation. This may be measured by reducing the amplifier so that the white radiation portion of the spectrum is selected by the voltage window (Fig. 5.21). This energy region must not contain any spectral lines and must be set well under the mean energy of the probe electrons. The signal entering the ratemeter is then collected for the preset time of the analysis. This measurement is of value for monitoring the specimen mass thickness (§ 5.3.4f). Its use in quantitative determinations is described in Chapter 6.

A convenient way of determining a suitable energy band for measurement of white radiation for instruments being operated at 30 keV or higher is to employ a pure silver standard (a disc of the metal will do, or a piece of silver wire, or even silver oxide powder). When the silver standard is irradiated the X-rays produced will be displayed as a sharp band of pulses on the CRO attached to the GFPC. The voltage window of the PHA is set to collect these pulses as described in § 5.3.5a. The voltage window is then displaced to an energy region slightly lower than that of the AgK_α line by changing the amplifier or the HT (high tension, or high voltage) to the GFPC. A plot of ratemeter signal against HT will show when the voltage window has moved away from the peak position and is measuring a part of the flat background or white radiation. The energy band near to the AgK_α line is a convenient one because there are few characteristic lines other than those of rare earth elements in that region. The operator should check on the X-ray lines by reference to X-ray tables (Bearden 1964).

White radiation may be measured with the GFPC of the crystal spectrometer also, since the background measured to the side of a characteristic line is truly a narrow band of continuum. This measurement is rarely sufficient, however, because of the inherently very low background produced by spectrometers compared to the GFPC or SSD.

5.3.6 Spurious X-ray lines

A number of X-ray lines may appear in the spectrum to confuse the analysis, the most important of these being *escape peaks* and, with crystal spectrometers, higher order diffracted lines.

5.3.6a Escape peaks

When an X-ray photon enters the X-ray detector (GFPC or SSD) it may cause a fluorescence of the detector material. For example, a CuK_α photon (8.047 keV) entering the GFPC may, in addition to producing ion pairs (§ 3.2.1b), also cause the inner orbital electron of a gas molecule (say of argon) to be ejected, thus producing a K_α emission from argon. The energy of the original CuK_α photon is thus reduced by an amount equivalent to the critical ionisation potential of the argon K_α shell (2.96 keV). The result is a peak in the spectrum corresponding to an X-ray energy of 5.09 keV, which is 2.96 keV lower than the main CuK_α line.

In an SSD the Si layer in the detector may be caused to fluoresce by incoming X-ray photons so that, for example, a high intensity CuK_α line produces a spurious peak at an energy E_{Cu}–E_c (Si), or 8.047–1.83 = 6.22 keV, where E_c (Si) is the critical ionisation potential of the SiK_α shell.

Pulses appearing in the escape peak position of the spectrum must be added to the main peak for proper representation of the incident X-ray intensity. Errors in intensity measurements due to escape peaks are generally much less than 10%.

5.3.6b *High order diffractions*

The equation governing the diffraction of X-rays by the spectrometer crystal, given in § 3.2.1, is $n\lambda = 2\text{d} \sin \theta$, where n is an integer. The most intense diffraction occurs when $n = 1$, but further strong reflection of X-rays with wavelength λ occurs at values of θ corresponding to $n = 2, 3$ etc. Thus there will be positions of the diffracting crystal at which spurious X-ray lines may occur that will confuse the analysis. A typical example of this is found when analysing a specimen on a support, such as copper, giving rise to a high intensity CuK_α line. The main CuK_α line will be found at a wavelength of 0.154 nm, but a further line will be detected at a wavelength of 0.308 nm which could well be confused with the K_α line of Sc (0.303 nm) or the K_β line of Ca (0.309 nm). Attention should be given to the position of these peaks in the spectrum during an analysis when one or two elements are producing high intensity lines. The problem does not occur when using a non-dispersive SSD.

References

Agar, A. W., R. H. Alderson and D. Chescoe (1974), Principles and practice of electron microscope operation, in: Practical methods in electron microscopy, Vol. 2, A. M. Glauert, ed. (North-Holland, Amsterdam).

Bahr, G. F., F. B. Johnson and E. Zeitler (1965), The elementary composition of organic objects after electron irradiation, Lab. Invest. *14*, 377.

Bearden, J. A. (1964), X-ray wavelengths, N.Y. 0-10586, US Atomic Energy Commission, Oak Ridge, Tennessee, USA.

Borom, M. P. and R. J. Hanneman (1966), Local composition changes in alkali silicate glasses during electron microprobe analysis, General Electric Technical Information Series, No. 66-C-484.

Bowman, M. J. and D. I. Hardie (1972), Cathode cap for the EM 6 microscope suitable for use with pointed filaments, J. Sci. Inst. (J. Phys. E) *1*, 9.

Bradley, D. E. (1965), The preparation of specimen support films, in: Techniques for electron microscopy, D. Kay, ed. (Blackwells, Oxford).

Chandler, J. A. (1973), Recent developments in analytical electron microscopy, J. Microscopy *98*, 359.

Dobb, M. G. (1972), The effect of cooling on the diffraction pattern of beam-sensitive specimens, Proc. 5th Eur. Reg. Conf. Electron Microscopy, Manchester, p. 564.

Fitzgerald, R. (1964), Beam stability in the electron probe microanalyser, in: Advances in X-ray analysis, Vol. 7, W. Mueller, G. Mallett and M. Fay, eds (Plenum Press, New York), p. 369.

Glaeser, R. M., V. E. Cosslett and U. Valdrè (1971), Low temperature electron microscopy: radiation damage in crystalline biological materials, J. Microscopie *12*, 133.

Goldstein, J. I. and H. Yakowitz (1975), Practical scanning electron microscopy; electron and ion microprobe analysis (Plenum Press, New York).
Green, M. and V. E. Cosslett (1968), Measurement of K, L, M shell X-ray production efficiencies, J. Phys. D., Applied Physics, JPDBA, Vol. 1, Series 2, 425.
Grubb, D. T. and A. Keller (1972), Beam-induced radiation damage in polymers and its effect on the image formed in the electron microscope, Proc. 5th Eur. Reg. Conf. Electron Microscopy, Manchester, p. 554.
Hall, T. A. (1971), The microprobe assay of chemical elements, in: Physical techniques in biomedical research, Vol. 1A, 2nd Edn. G. Oster, ed. (Academic Press, New York).
Hall, T. A. and B. L. Gupta (1974), Beam induced loss of organic mass under electron microprobe conditions, J. Microscopy *100*, 177.
Hall, T. A. and P. Werba (1969), The measurement of total mass per unit area and elemental weight fractions along line scans in thin specimens, Proc. 5th Int. Congr, X-ray Optics and X-ray Microanalysis, Tubingen, G. Mollenstadt and K. H. Gaukler, eds (Springer Verlag, Berlin).
Hearle, J. W. S., J. I. Sparrow and P. M. Cross (1972), The use of the scanning electron microscope, (Pergamon Press, Oxford.).
Hodson, S. and J. Marshall (1970), Tissue sodium and potassium: direct detection in the electron microscope, Experientia *26*, 1283.
Höhling, H. J., T. A. Hall, W. Kriz, A. P. V. Rosenstiel, J. Schermann and V. Zessack (1971), Loss of mass in biological specimens during electron probe X-ray microanalysis, Proc. Int. Symp. Med. Tech. in Physiol. Sci., Munich, p. 135.
Kritzinger, S. and E. Ronander (1974), Local beam heating in metallic electron microscope specimens, J. Microscopy *102*, 117.
Oatley, C. W. (1972), The scanning electron microscope, Part 1, The instrument (Cambridge University Press).
Reed, S. J. B. (1968), Probe current stability in electron probe microanalysis, J. Phys. E., Sci. Instruments, JPSIA, Vol. 1, Series 2, 136.
Reimer, L. (1965), Irradiation changes in organic and inorganic objects, Lab. Invest. *14*, 344.
Saubermann, A. J. and P. Echlin (1975), The preparation, examination and analysis of frozen hydrated sections by scanning transmission electron microscopy and X-ray microanalysis, J. Microscopy *105*, 155.
Schamber, F. H. (1973), A new technique for deconvolution of complex X-ray energy spectra, Proc. 8th Nat. Conf. Electron Probe Analysis, paper no. 85A.
Stenn, K. S. and G. F. Bahr (1970), A study of mass loss and product formation after irradiation of some dry amino acids, peptides, polypeptides and proteins with an electron beam of low current density, J. Histochem. Cytochem. *18*, 574.
Thach, R. E. and S. S. Thach (1971), Damage to biological samples causes by the electron beam during electron microscopy, Biophys. J. *11*, 204.
Thornton, P. R. (1968), Scanning electron microscopy (Chapman & Hall, London).

Chapter 6

Quantitative X-ray microanalysis

X-ray microanalysis is in the first place a very rapid method of qualitatively deciding what elements are present in the region of the specimen being examined. To advance from this point to determine quantitative concentrations of each element requires a little determination in handling mathematical formulae. In addition the operator must be familiar with all the practical pitfalls in each operation.

The complexity of the measurements made will vary depending on what information is required from the specimen. The simplest analysis is the study of the relative distribution of an element, or elements, across a specimen in either the thick or thin form, without regard to absolute concentrations. In thick specimens this actually involves plotting concentrations, but in thin specimens the specimen mass thickness must be taken into account for the determination of concentrations. If absolute concentrations are required then it is usually necessary to use standards; that is, samples with known concentrations of an element (§ 4.5).

Quantitative X-ray microanalysis as applied to very thick specimens can be extremely complex and is not discussed in detail here. For a thorough description of bulk specimen procedures the reader is referred to the references listed under Further Reading at the end of this chapter.

In order to understand the basic principles of performing simple quantitative analyses, however, it is useful to first discuss the effects of electron-specimen interaction in bulk specimens with reference to the generation of X-rays.

6.1 *Quantitative correction procedures – the ZAF method (for bulk specimens)*

In order to produce an X-ray photon the incident electron must have sufficient energy to ionise the atom concerned (§ 2.4) and this energy is called the critical excitation potential E_c, or the absorption edge energy. Electrons with energy much greater than E_c may produce several ionisations and so progressively lose energy, or they may be elastically scattered, whereby they lose no energy but may change direction. In addition, they may be decelerated in the nuclear field of an atom to produce white radiation (§ 2.4.4).

The concentration, C_x, of an element x in a bulk sample compared with a standard of the pure element is given simply by $C_x = I_x/I_{x_s} \cdot C_s$, or $C_x = I_x/I_{x_s}$ since $C_s = 1$, where I_x and I_{x_s} are the intensities of characteristic X-ray emission from the specimen and the standard respectively. The intensity of the X-rays from the standard is always the same, given the same analytical conditions, and so $C_x = k\, I_x$ where k is a constant.

As the electrons enter a very thick sample (of greater thickness than the electron path) a number of processes occur which affect the production and collection of X-rays from the sample. The equation for calculation of elemental concentration is then modified to

$$C_x = \frac{k\, I_x}{C_Z\, C_A\, C_F}$$

where I_x is the X-ray intensity from the sample, k is a constant again depending on the intensity obtained from a standard, and C_Z, C_A and C_F are correction factors. These correction factors are required to account for the effects of atomic number, X-ray absorption, and X-ray fluorescence respectively. The three correction factors make up what is known as the ZAF correction procedure for use in microanalysis of bulk specimens (Philibert and Tixier 1968). Such correction procedures are often very complex mathematically and are frequently carried out on a computer (Warner and Coleman 1974; Yakowitz 1975). For the types of analyses discussed in this book, that is for thin specimens examined in the electron microscope, they are not generally required, however, but it is important to understand the underlying theories. The correction factors are considered here in turn.

6.1.1 *The atomic number effect* (C_Z)

Electrons entering the surface of a specimen both penetrate the specimen and are scattered by it (Fig. 6.1). A complex calculation taking into account the random nature of these interactions (the Monte-Carlo method) can be used in computer correction procedures (Green 1963). The most widely used of such methods is the one described by Philibert and Tixier (1968).

The effects of these processes on X-ray emission can be considered more simply in terms of the two factors R and S. The electrons which are backscattered from the specimen surface do not contribute to X-ray production. The fraction of the incident electrons which enters the sample and remains within it is called R. Specimens of low atomic number (such as biological specimens) produce less backscattering and consequently a higher value for R.

The electrons which penetrate the specimen may cause ionisations, producing X-rays, or they may be scattered within it (§ 2.4). The production of X-rays by ionisation depends on the critical ionisation potential of the

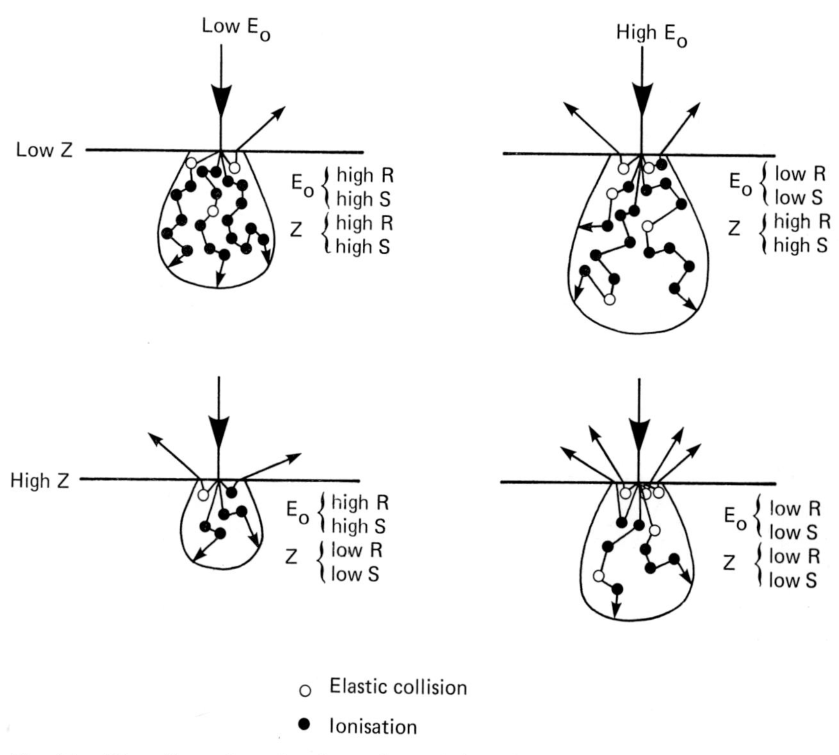

Fig. 6.1. The effect of accelerating voltage (E_0) and atomic number (Z) on electron penetration and ionisation. The fraction of the incident electrons remaining in the specimen (R) and the stopping power per unit mass of the specimen (S) vary together.

specimen and thus on its composition. Elements of low atomic number (Z) have lower critical ionisation potentials, are more easily ionised, and have greater stopping power per unit mass. The stopping power of the specimen is called S, and is higher for elements of low atomic number.

The initial energy (E_0) of the electrons affects the values of R and S also. Higher energy electrons may more readily be backscattered and escape from a sample, producing a lower value for R, while the stopping power of the sample may be reduced, producing a lower value for S.

Thus we have R and S varying inversely with both E_0 and Z (although not smoothly). This is fortunate since the correction to be applied depends on the ratio of R and S, i.e.

$$C_Z = R/S$$

Thus the two factors tend to cancel each other out as both the atomic number and electron accelerating voltage vary (Fig. 6.1). Although the correction for C_Z is minimised in this way, it still may represent an important part of the correction procedure and is included in computer correction programmes.

6.1.2 *The absorption effect* (C_A)

As shown in § 2.4.3, X-rays travelling through a material of density ρ suffer absorption to the degree

$$\frac{I_t}{I_0} = e^{\frac{-\mu}{\rho} \cdot \rho x}$$

where I_t is the transmitted X-ray intensity, I_0 the original intensity, μ/ρ is the mass absorption coefficient, and x is the distance travelled by the X-rays. The value of μ/ρ depends on the X-ray energy as well as on the composition of the specimen. For a specimen composed of a number of elements the value for μ/ρ is given by the sum of the individual absorption coefficients multiplied by their mass fractions (§ 6.2.1).

X-rays are generated at different depths within a specimen and hence must travel through different distances to leave the sample. They are thus absorbed to different degrees. The correction for this absorption, C_A, takes into account the shape of the volume of specimen producing the X-rays (§ 3.1.1); the angle at which the X-rays are collected from the specimen surface; the angle at which the incident electron beam enters the specimen surface; and the composition of the specimen.

6.1.3 The fluorescence effect (C_F)

The absorption of X-rays travelling through the sample occurs when the X-ray photons give up their energy in ionising other atoms (§ 2.4.2). Thus if an X-ray photon of element Z_1 has an emission energy which is slightly more than the energy required to ionise element Z_2 (the absorption edge, or critical excitation potential), that photon is very likely to be absorbed, and element Z_2 will emit X-ray photons. Thus if element Z_2 is being measured, its intensity will be raised and its apparent concentration artificially increased. The degree of enhancement depends on the relative excitation probability (§ 2.4.1), the fluorescent yield (§ 2.4.1), and the depth of electron penetration and X-ray production within the sample (which is a function of electron energy).

In addition X-ray enhancement may occur due to fluorescence by the white radiation, the effect being greatest in specimens of high atomic number which produce the greatest bremsstrahlung. It may be largely ignored with thin biological samples, but becomes more serious as thickness increases (Hall 1971).

The combination of these parameters makes up the correction factor for fluorescence, C_Z.

6.1.4 The standard constant (k)

The equation for correction of elemental concentration

$$C_x = k \cdot \frac{I_x}{C_Z\ C_A\ C_F}$$

has a constant (k) which depends upon data obtained from a standard. Pure element standards are usually used for each element being investigated, and when no pure standards are available for certain elements the intensity ratio between that element and another, obtainable in the pure form, is used. This alternative standard is usually of an element close in the periodic table to the one being examined in order to simulate the appropriate C_Z and C_A effects. For pure elements fluorescence does not occur, of course, (except a small contribution from bremsstrahlung) and is ignored.

Thus, performing the correction procedures for C_Z and C_A on a pure element standard allows the constant k to be calculated and provides all the data necessary to calculate the concentration in the specimen.

The majority of analyses of bulk material are performed in this way. The operator needs only to measure the intensity from a standard and then, using identical conditions of analysis on the specimen, obtain a value for the intensity of characteristic radiation. Inserting this data, together with information regarding the electron accelerating voltage, elemental composition, specimen angle, X-ray take off angle, etc, is sufficient for the computer to calculate elemental concentrations. The correction procedures and the computerised methods for use in X-ray analytical studies have been largely developed for metallurgical and mineralogical bulk specimens. The analysis of bulk specimens of biological material poses certain problems in the use of pure standards, however. Hall (1968, 1971) has considered the theory of quantitative analysis of biological bulk specimens and Warner and Coleman (1974) describe some computer programmes for conventional analytical methods.

An alternative form of analysis of bulk biological specimens is described by Cobet (1972). He established the relationship

$$C_x = k \cdot \frac{I_x}{W}$$

where I_x is the characteristic intensity for element x, W is the continuum produced from the specimen and measured near to the energy of I_x, and k is a constant, again derived from a standard. This formula minimises the need to use correction factors C_Z, C_A and C_F and can greatly simplify quantitative analyses of biological specimens.

6.2 *Thin specimen analysis*

Most of the correction procedures outlined in § 6.1.1 do not apply to thin specimens because of the lack of backscatter and electron energy loss. Absorption and fluorescence may also be ignored when considering elements of atomic number > 11. The most important factor in determining the X-ray generation in a thin specimen is the electron accelerating voltage E_0, since the relative ionisation probability (or ionisation cross section) is given as $Q = (1/E_0\,E_c)\log(E_0/E_c)$ (see § 2.4.1).

This reaches a maximum for values of $E_0 \simeq 2.7\,E_c$, but because of increased gun brightness at higher accelerating voltages it is usual to perform analyses in the TEM at 80 or 100 kV, and in the SEM at around 30 kV (i.e. at the maximum kV available) (§ 5.2.1) and E_0 is thus considerably greater than $2.7\,E_c$.

There are two main methods of performing quantitative analyses on thin specimens, and these are considered in turn.

6.2.1 *The continuum method*

The characteristic X-ray intensity is proportional to the mass per unit area of the element being irradiated in a thin specimen, and the white radiation intensity is proportional to the mass per unit area of the whole specimen in the region being irradiated (Hall 1971). The white radiation thus depends on the mean atomic number of the specimen in the region being irradiated (Hall and Werba 1971; Hall et al. 1973; Hall 1974).

Thus the relative concentration (mass fraction) of an element in a thin specimen may be represented by the ratio

$$C_x = K \cdot \frac{I_x}{W}$$

where I_x is the characteristic X-ray intensity and W is the white radiation from the specimen. In contrast to the similar equation for bulk specimen analysis (§ 6.1.4), W does not have to be measured near the energy of I_x. Since no corrections are necessary for electron backscatter or penetration, and absorption and fluorescence may be considered negligible in thin specimens, the equation is not complicated by correction procedures such as the ZAF routines.

If a standard is used (§ 4.5.2), the value for K can be readily determined as $K = (I_x/W)_{\mathrm{standard}}$. All subsequent concentrations are calculated by the simultaneous measurement of I_x and W. The method works for both thin and ultrathin specimens and is widely used for microanalysis in the SEM and TEM. There are, however, some limitations to the usefulness of this method for both types of specimen.

A primary electron entering a bulk specimen causes ionisations until its energy falls below the critical excitation potential E_c (§ 3.1.1). Continuous radiation is still produced, however, until the electron energy falls to zero. In ultrathin specimens the energy loss of transmitted electrons is very small and no corrections are necessary for changes in X-ray production due to energy loss. In addition there is no backscatter and absorption is negligible. As the thickness increases, however, it eventually enters the range where these factors become significant.

The maximum thickness for which the simple equation for C_x applies is that beyond which the ratio of characteristic to white radiation no longer remains constant. This thickness depends on the density of the specimen and on the accelerating voltage. For a biological specimen it is likely to be in the region of 10 μm when using an accelerating voltage of 100 kV, and to be less at lower voltages. For a specimen of greater density, the I_x/W relationship ceases to be linear for thicknesses lower than 10 μm, the change from linearity occurring earlier for the lighter elements. Sweeney et al. (1960) have determined the range of linearity for evaporated metal films. For a satisfactory analysis using the continuum method, the standard must also fall within these thickness limitations.

Further limitations on the accuracy of the method lie in the nature of the specimen being analysed. If biological specimens are embedded in resin, then elemental concentrations determined by this method refer only to the specimen being examined and may not represent those of the original sample. This is because the major part of the white radiation comes from the resin itself which, of course, is not present in the original tissue. If it may be assumed that resin replaces the tissue water content (sometimes a good approximation) then *in vivo* concentrations can be calculated.

For this reason it is often advantageous to work with frozen sections which may be examined in the freeze-dried condition (§ 4.4.2c) or in the frozen-hydrated form (§ 4.4.1c). Such specimens are, at the present time, more likely to be examined in the SEM. However, care needs to be taken over the interpretation of data from frozen sections analysed in this way because of the possible collapse of tissue after freeze-drying (Hall 1971).

The simplified formula above for C_x is correct when a standard is employed having an almost identical composition (mean atomic number) to the specimen. For biological work this means using organic standards; for metallurgical or mineralogical work, non-organic thin standards are employed.

Strictly, the equation developed by Hall (1971) for determination of concentrations is given by

$$C_x = \frac{(I_x/W)_{\text{sp}}}{(I_x/W)_{\text{st}}} \left[\frac{N_x}{\sum NZ^2}\right]_{\text{st}} \cdot A_x \cdot \overline{(Z^2/A)}_{\text{sp}}$$

where C_x is the concentration of element x to be determined;

$$(I_x/W)_{\text{sp}} \quad \text{and} \quad (I_x/W)_{\text{st}}$$

are the ratios of peak intensity to white radiation for the specimen and standard respectively; N_x is the number of atoms of element x per unit volume in the standard and ΣNZ^2 is the summation of the number of atoms of each element multiplied by the square of its atomic number in the standard (of known composition); A_x is the atomic weight of the element x, and $\overline{(Z^2/A)}_{sp}$ is the mean of the ratio of the square of the atomic number divided by the atomic weight for each element in the specimen being examined.

It has been found (Hall 1971) that for an organic specimen of typical composition 7%H, 50%C, 25%O, 16%N and 2% (S + P), the value of $\overline{(Z^2/A)}$ is 3.28. For non-biological specimens certain assumptions must be made about the composition of the specimen initially (from a semi-quantitative analysis of a complete spectrum) and the relationship calculated iteratively by computer.

A_x is known, and from the standard

$$\left[\frac{N_x}{\sum NZ^2}\right]_{st}$$

can be easily calculated. Therefore all that needs to be measured is the ratio (I_x/W) for both specimen and standard. The equation allows the use of standards which are not exactly similar to the specimen being examined, although, since the equation is an approximation from a more complex form (Hall 1971), standards should at least approximate to the specimen (Hall et al. 1973; Spurr 1975). For biological specimens this often means using organic matrices or salts of low average atomic number.

Concentrations are calculated using this formula either by hand or by computer methods. A number of suppliers of energy dispersive analysers with computer facilities (see Appendix) provide programmes for using such an expression. A useful worked example using this formula for analysis of a biological thin section is given by Hall et al. (1973).

The method may also be employed for non-biological thin specimens. It is important when using standards, however, to have some knowledge of the average composition of the standard (§ 4.5.2). The use of the continuum method for application to thin metal films has been described theoretically by Marshall and Hall (1968) and by Philibert et al. (1970).

For specimens of intermediate thickness, where the linearity of emission with thickness no longer holds, the ZAF corrections begin to apply for fluorescence, absorption, penetration and backscatter (§ 6.1). The equations then become more complex and are usually solved by computer techniques (Warner and Coleman 1974; Yakowitz 1975).

6.2.1a Spurious contributions to the X-ray signal

Even with a well-focused electron beam there will be some electron scatter from the specimen producing spurious radiation from surrounding parts of

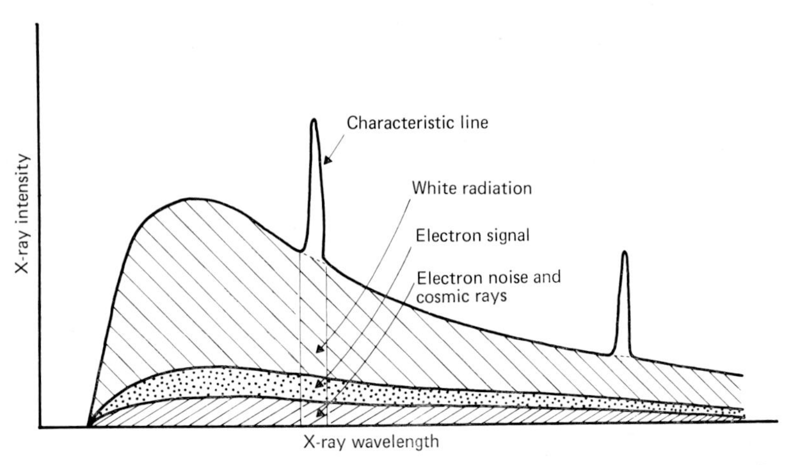

Fig. 6.2. The contributions to background radiation beneath the characteristic lines. White radiation, caused by electron interaction with atomic nuclei, forms the bulk of the background.

the microscope or specimen support (Fig. 5.1). This spurious radiation may contain characteristic X-ray lines which interfere with the spectrum emitted by the specimen (see § 4.1.1 and § 5.1.4), and it will contain white radiation due to electron interaction with nuclei which effectively introduces a background signal (Fig. 6.2). The use of light elements such as carbon or beryllium in the specimen environment, and grids of large mesh size partly overcomes this problem (Chandler 1973; Cooke and Duncumb 1966).

When analysing biological materials, the presence of heavy metal stains can be troublesome (§ 4.4.2a) and can make the determination of specimen mass thickness very difficult. The various contributions to the background, the bulk of which is white radiation due to electron interaction within the specimen, are shown in Fig. 6.2. To demonstrate the method for dealing with spurious signals for quantitative purposes an example of a biological specimen is used. The problems are common to both ultrathin specimens examined in the TEM and thin specimens analysed in the SEM, for biological and for non-biological specimens.

When the electron beam strikes the specimen, some electrons are scattered towards the X-ray window and detector, others interact with grid bars and surrounding metal parts. The total X-ray signal entering the detector is thus given by:

$$P_x = I_x + \Delta W_{\mathrm{Stain}_1} + \Delta W_{\mathrm{Stain}_2} + \Delta W_E + \Delta W_S + \Delta W_{\mathrm{Cu}} + \Delta W_W + e$$

where ΔW represents that fraction of the white radiation under the characteristic line.

and I_x is the true characteristic signal from element x. $\Delta W_{\text{Stain}_1}$ is the white radiation due to heavy metal stain 1; $\Delta W_{\text{Stain}_2}$ is the white radiation due to heavy metal stain 2, etc.; ΔW_E is the white radiation due to the embedding medium containing other endogenous elements; ΔW_S is the white radiation due to the support film (carbon, formvar, etc); ΔW_{Cu} is the white radiation due to electrons interacting with the grid bars, pole pieces, etc. (in this example copper, but other elements may occur); ΔW_W is the white radiation due to electrons interacting with the X-ray window; and e is the signal due to scattered electrons directly entering the detector.

If the background (b_x), comprising these radiations, is measured (at an energy or wavelength slightly away from the characteristic line), then the true characteristic line value is given by

$$I_x = P_x - b_x$$

The characteristic signal from the region surrounding the specimen (the support film) is then measured in the same way and is given by

$$I_x^1 = P_x^1 - b_x^1$$

To obtain the relative concentration of an element from place to place on the specimen the value for each characteristic signal is divided by the mass thickness of the specimen as monitored by the total white radiation collected from the specimen region irradiated. The energy dispersive detector or gas-flow proportional counter can be used to measure this signal (W_T) which is given by

$$W_T = W_E + (W_{\text{Stain}_1} + W_{\text{Stain}_2} + W_s + W_{\text{Cu}} + W_W + e')$$

where W_T is the white radiation measured over a band of the energy spectrum containing no characteristic lines, and W_{Stain_1}, W_{Stain_2}, W_s, W_{Cu} and W_W are the white radiation contributions from specimen stains (which vary from place to place), support film, and electron interaction with copper grid bars and X-ray window, and e' is again the direct electron signal entering the detector (different from e, above, and varying across the specimen).

W_E is the white radiation from the selected region of the section and its value is found by measuring the other factors separately.

W_{Stain_1} can be estimated from the relationship $W_{\text{Stain}_1} = k_1 I_1$ where k_1 is a constant (previously determined on a standard) and I_1 is the characteristic reading for the element contained in the stain as measured on the spectrometer or the SSD. The same is true for W_{Stain_2}.

For W_{Cu} the relationship is $W_{\text{Cu}} = k_{\text{Cu}}\, I_{\text{Cu}}$ where k_{Cu} is determined by using a pure copper specimen as a standard. This is generally the largest contribution to spurious white radiation and can be reduced by using light element specimen supports (§ 4.1) and grids of large mesh size.

W_W depends on the thickness of the X-ray window but is usually small enough to be neglected. W_s is measured on the supporting film alone.

The value for e' varies from place to place on the specimen according to its scattering power (amount of stain, etc.). It can become a significant part of the total white radiation and must be minimised by correct filtering techniques (§ 5.2.6c and § 5.2.6e).

A satisfactory arrangement for making these corrections consists of an energy dispersive detector or GFPC employed to measure white radiation with a thick window electron filter, and with the spectrometers or SSD used to measure all the characteristic lines in sequence.

Thus the value for relative concentration, C_x, of element x in the specimen is finally given by

$$C_x = \frac{K\,[I_x - I'_x]}{W_E} = \frac{K\,[(P_x - b_x) - (P^1_x - b^1_x)]}{[W_T - (W_{\mathrm{Stain}_1} + W_{\mathrm{Stain}_2} + W_s + W_{\mathrm{Cu}})]}$$

$$= \frac{K\,[(P_x - b_x) - (P^1_x - b^1_x)]}{[W_T - (k_1\,I_{\mathrm{Stain}_1} + k_2\,I_{\mathrm{Stain}_2} + W_s + k_{\mathrm{Cu}}\,I_{\mathrm{Cu}})]}$$

The value for C_x can be compared with that for a standard specimen having the element in known concentration, to obtain the value for K and hence to calculate elemental concentrations.

The relationship can be greatly simplified if the copper signal is reduced, e.g. by using beryllium or carbon grids, by careful design of the specimen stage and by examining the specimen unstained and unsupported. Then

$$C_x = K \cdot \frac{I_x}{W_T}$$

although a small white radiation signal may be contributed even by carbon or beryllium grids.

Hall (1971) and Hall et al. (1973) give a fuller treatment of the processes involved in producing the total white radiation W_T.

6.2.2 *The ratio method*

6.2.2a *The ratio method – using thin standards*

If it is not necessary or expedient to measure absolute elemental concentrations, then a simple calculation of the ratios of two elements across a thin specimen may be sufficient. If the absolute concentration of one of these elements is known in the specimen it may act as an internal standard and allow real concentrations to be measured.

The relative elemental atomic concentrations (C_1 and C_2) in a specimen can be directly calculated from the ratio of their X-ray intensities (I_1 and I_2):

$$\frac{C_1}{C_2} = k_s \frac{I_1}{I_2}$$

where k_s is a constant that accounts for the *relative detection efficiency* of the whole analysis system (§ 6.3, § 3.3.2) for the two elements. k_s is easily deduced from a thin standard having a well known proportion of the elements 1 and 2 (§ 4.5.2). I_1 and I_2 may be measured on an energy dispersive detector or with crystal spectrometers.

If the relative detection efficiency of all the elements being examined is known by prior calibration with appropriate thin standards, then the relative concentrations of a number of elements in the specimen may be compared:

$$C_1 : C_2 : C_3 = \frac{I_1}{k_1} : \frac{I_2}{k_2} : \frac{I_3}{k_3}$$

where k_1, k_2, and k_3 are constants of proportionality for collection efficiency of each element. The method needs no other standards and does not depend on measuring the white radiation. The following example of an analysis of individual mineral particles may help to clarify the application of the method.

(i) A standard is prepared from a composite slurry of SiO_2 and $Al(OH)_3$ as described in § 4.5.2a.

(ii) Analysis of a thinly dispersed layer of this standard is performed to give a ratio of Al : Si X-ray intensities, say R_{st}. This ratio is determined by analysing a large number of particles simultaneously to eliminate local fluctuations in particle composition.

(iii) Analysis is then performed on individual particles of the specimen, and values for the ratio of Al : Si X-ray intensities determined, say R_{sp}.

(iv) If the atomic ratio of Al : Si in the standard is 1 : 1, the atomic ratio of these two elements in a particle in the specimen is given by R_{sp}/R_{st}.

In many types of analysis the specimen may act as its own internal standard if the overall elemental ratio is known and it is desired to monitor local ratios across the specimen.

It is usual to calibrate the X-ray detector for relative collection efficiency for a wide range of elements so that any other elemental ratios may be determined in a similar way. Composite standards (see Appendix) can be purchased and are often used for this type of calibration, but a simple method for preparing and using calibration standards is discussed in § 6.3.

6.2.2b *The ratio method – using no standards*

The method for determining elemental ratios described above involves a knowledge of the constant k_s which is derived from a thin standard. Russ (1974) has shown that this constant may also be derived from theory without using standards when a SSD is employed.

The method relates mass fraction C_x to characteristic intensity I_x in the form

$$C_1 : C_2 = \frac{I_{x1}}{k_1} : \frac{I_{x2}}{k_2}$$

where $k = Q\,(WR)\,T/A$

Q = ionisation cross section (§ 2.4.1)

(WR) = fluorescent yield and relative line intensity (§ 2.4.1)

T = efficiency of detector system (§ 6.3, § 3.3.2)

and A = atomic weight.

The value of k may be determined from thin standards as described before or by calculation since all the factors except T can be found from tables. T depends on the X-ray energy, angle of X-ray collection and thickness of window material between the specimen and the detector. It may be deduced as shown in § 6.3.

The formula for $C_1 : C_2$ is incorporated into a computer programme and the values for the different factors entered before the analysis. When I_{x_1} and I_{x_2} are obtained from analysis of the specimen, the elemental mass ratio is then calculated automatically.

6.2.2c *The ratio method – using bulk standards*

The ratio of the concentrations of two elements may be determined in a thin specimen in a similar fashion to that described in § 6.2.2a, but instead of thin standards, bulk standards may be employed. Duncumb (1968) calculated that the measured ratio I_x/I_y of the X-ray intensities of two elements x and y in a metal foil (or mineral specimen) is related to the mass concentration ratio C_x/C_y as follows:

$$\frac{C_x}{C_y} = \frac{I_x}{I_y} \cdot \frac{Q_{oy}}{Q_{ox}} \cdot \frac{I'_{(y)}}{I'_{(x)}}$$

where Q_{ox} and Q_{oy} are the ionisation cross-sections for electrons with the incident beam energy E_o, and $I'_{(x)}$ and $I'_{(y)}$ are the X-ray intensities from pure bulk standards corrected by the ZAF procedure (§ 6.1). This equation was proposed for thin metal foils examined in the SEM at voltages up to 30 kV, and was later verified (Jacobs and Baborovska 1972) for foils examined in the TEM/EMMA at energies up to 100 kV. For the equation to be valid the foil must have a thickness less than that causing a change in linearity of I_x/I_y. Jacobs and Baborovska found that for AuL_α emission in a 25% Cu – 75% Au foil the maximum allowable thickness was in the region of 0.3 μm before electron penetration effects occurred.

The equation may be employed in a simple computer programme and the X-ray data fed in to provide readily calculated elemental ratios.

6.3 *Calibration of the X-ray detector – relative efficiency*

The method described for determining elemental ratios may be extended to allow calibration of the X-ray detection system for all elements in the detectable range.

A typical response curve for an SSD collecting X-rays of varying energy but of similar intensity is shown in § 3.3.2. This curve, which is usually supplied by the company manufacturing the detector, does not, however, represent the efficiency of detection of *the whole analysis system*. Factors to be taken into account include the X-ray yield in the specimen (§ 2.4.1); absorption in the X-ray window material (§ 5.2.7b); possible absorption in the medium (air or vacuum) before the X-ray detector (§ 5.2.7c); absorption in the beryllium window of the detector itself (§ 5.2.7e); and efficiency of collection by the detector (Fig. 3.12). The efficiency of X-ray detection varies in the GFPC of the crystal spectrometer in a similar manner.

To obtain a true response curve for the system and operating conditions selected, a number of standards must be chosen that include a variety of elements in combination for calibration one against another. Ordinary laboratory reagents may be employed in a similar manner. For example, a number of sulphate or sulphide salts, prepared as thin specimens by drying down from suspension (§ 4.3.2) allow other elements to be related to sulphur, as follows:

(i) Make up thin specimens, on carbon-coated grids, of the following salts: NaCl, $MgSO_4$, Al_2SiO_3, $CaPO_4$, $CaCl_2$, $CaSO_4$, $FeSO_4$, $CuSO_4$, $CuCl_2$, $ZnSO_4$.

Other similar salts may be made for the heavier elements. The salts

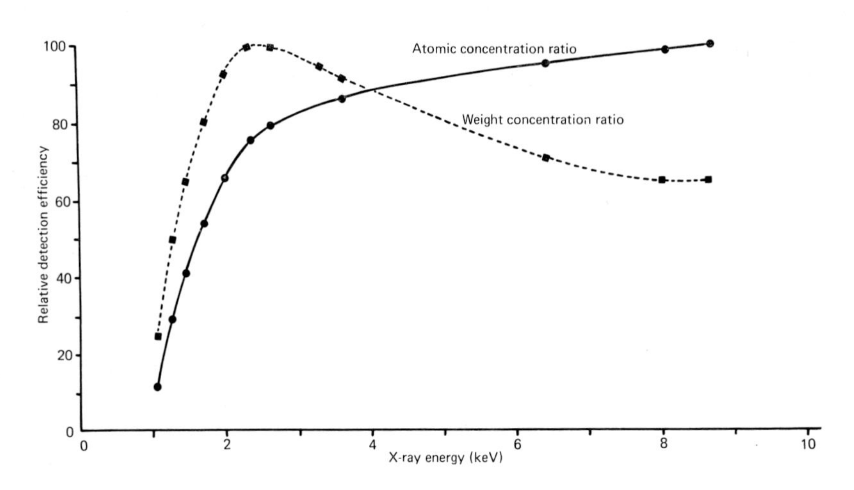

Fig. 6.3. Calibration curves for relative detection efficiency with varying X-ray energy.

used must be of high purity (Analar grade). Powdered minerals ground from bulk specimens, with well-known elemental compositions (§ 4.5.2a), may also be used as standards. It is essential that the standard remains stable in the electron beam and this must be ascertained before it is used for calibration purposes. Carbon coating the standards will assist the stability.

(ii) Obtain X-ray signal ratios for each of the elemental pairs.

(iii) Normalise every elemental signal against the S signal (i.e. Na/S = Na/Cl × Cl/Cu × Cu/S).

(iv) Construct a curve of relative efficiency of detection such as that shown in Fig. 6.3, either for ratios of numbers of atoms or for mass ratios.

Whenever an analysis is performed for a number of elements this curve will allow corrections to be made for relative efficiencies. A similar curve is constructed for the crystal spectrometer for each crystal.

NB: The curve is dependent on operating conditions (kV in particular), and on the X-ray window thickness. The standards must be thin and stable in the electron beam (See Sweatman and Long 1969; Hall and Peters 1974; Cliff and Lorimer 1975).

NB: The calibration efficiency should be checked (§ 5.2.7b), every time the window in front of the X-ray detector is changed to ensure reproducible analyses.

6.3.1 *Absolute calibration*

In order to achieve absolute quantitation of any element a single standard may be prepared as described in § 4.5.2. This standard may then be used to fix the co-ordinates of the curve shown in Fig. 6.3.

For example, in the analysis of a thin specimen, a standard composed of a resin containing a fixed concentration of potassium (§ 4.5.2b) will provide a ratio of characteristic radiation to white radiation intensity, I_x/W. In the graph of Fig. 6.4 a 1% potassium standard gave an I_K/W ratio of 0.14. If it was required to analyse zinc in a biological specimen, the relative efficiency curve of Fig. 6.3 would indicate that

$$\frac{I_{Zn}}{I_K} = \frac{65}{95} = 0.68$$

Thus a 1% zinc standard would give a ratio of characteristic to white radiation of

$$\frac{I_{Zn}}{W} = 0.68 \times 0.14 = 0.095$$

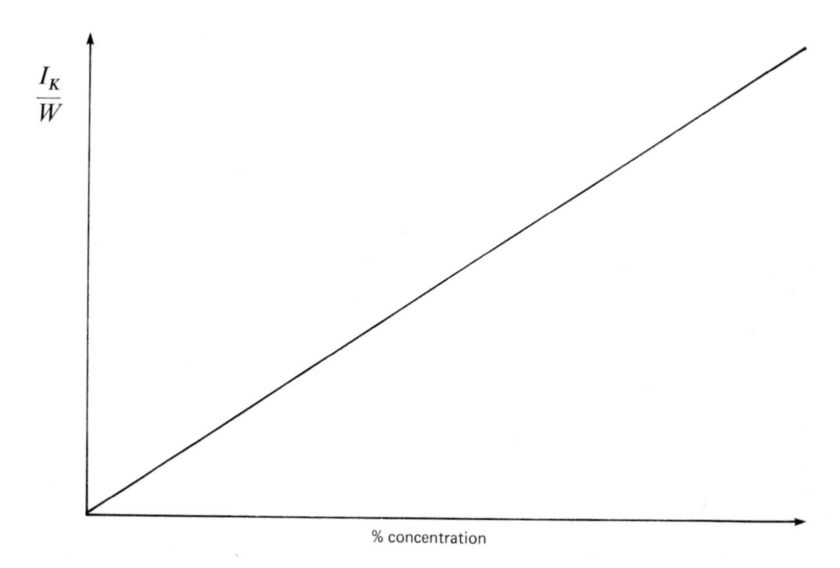

Fig. 6.4. The relationship between the ratio of characteristic to white radiation versus % concentration obtained from thin standards. The curve is linear and may be used to convert X-ray counts to elemental concentrations.

All successive values of I_{Zn}/W measured on the specimen being analysed would then give concentrations

$$C_{Zn} = \frac{(I_{Zn}/W) \times 1}{0.095} \%$$

The accuracy of this relationship depends on the limitations described in § 6.2.1.

Calibration curves for characteristic/white intensity versus % concentration (Fig. 6.4) can be drawn for any element, if the relative efficiency curves (Fig. 6.3) have been calculated. For metallurgical or mineralogical specimens a similar procedure can be adopted using the appropriate standards, but again allowing for the mean atomic number effect (§ 6.2.1).

Calibration curves can be constructed for the crystal spectrometer and SSD in a similar way.

6.4 *Some practical considerations*

6.4.1 *Contamination and specimen damage*

The problem of contamination (§ 5.1.3) has important consequences in the analysis of thin, and especially ultrathin, specimens. Effectively, in a less than ultraclean vacuum system, the electron beam causes the polymerisation of organic molecules adsorbed by the specimen. In time this contaminating layer (mostly of carbon) can build up to a level similar to, or even greater than, the specimen thickness itself. The effect on quantitative measurements is to drastically change the specimen mass thickness even during the period of analysis. The effect is worse when using lower beam currents since the specimen is locally less heated.

Good vacuum systems and liquid nitrogen-cooled anti-contaminators are the main lines of defence against contamination. Except for the very lightest elements ($Z < 11$) contamination has little effect on the characteristic radiation, and elemental ratios are unchanged.

The degree to which contamination affects the specimen mass thickness may be estimated by counting the white radiation over a period of time while irradiating the specimen (Fig. 6.5). If it increases during the analysis time then errors may be introduced into the calculations of elemental concentrations by the continuum method. In addition, deposition of contamination during the analysis also makes the successive measurement of peak and background with the crystal spectrometers very difficult. If this is a problem,

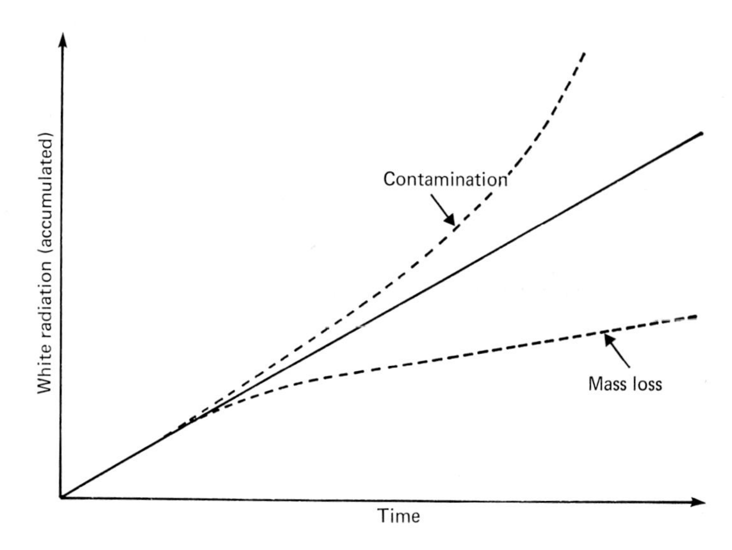

Fig. 6.5. The effect of contamination on white radiation measured on a thin specimen with time. Contamination may introduce serious errors in measurement of specimen mass thickness.

then the method of Hall and Werba (1969) for simultaneous measurement of peak and background may be helpful. In this technique the peak value is read with the spectrometer tuned to the appropriate spectral line, and another detector simultaneously measures the background. Any of the 3 types of detector may be employed. If the instrument is fitted with twin crystal spectrometers then the second spectrometer can be used. It must be ensured that background readings at a wavelength setting away from the characteristic line are similar, or at least may be compared, in the two spectrometers.

If an SSD or GFPC is also used then an energy band may be selected to read white radiation (§ 5.3.4f, § 5.3.5b). The background for the spectrometer reading will then be proportional to the white radiation. This ratio can be determined simply by tuning the crystal spectrometer to the background position and comparing the X-ray intensity with the white radiation from the non-dispersive detector.

In addition to contamination, the problem of specimen mass loss in the electron beam may be very important in quantitation. This loss is more likely to occur in biological samples and can be measured as described in § 5.1.2. The effect of mass loss on the white radiation is shown in Fig. 6.5. The effect of loss of an element in the electron beam as a result of radiation damage or heating effects (§ 5.1.2) will produce a similar fall-off of X-ray intensity for that element to that shown for the white radiation. Unfortunately much of the losses of mass or elemental content may occur in the early stages of irradiation, even within the first few milliseconds (Hall and Gupta 1974), and may be difficult to determine. Much work has yet to be done using cold stages to determine the effect of temperature on the degree of loss from biological specimens (§ 5.1.2). Thin standards with varying concentrations of chosen elements can be used to determine these losses and the resultant effects on quantitation (Chandler 1976).

6.4.2 *Count rates and sensitivity of analysis*

With most biological materials the concentration of the element being analysed is very low ($< 1\%$), and with ultrathin sections of embedded specimens or with very thin films, the actual mass of the element is very small indeed. This may also be true of some mineralogical specimens. As a consequence the count rates are usually extremely low and care has to be taken to achieve statistically meaningful results. In order to raise the count rate the beam current and counting time can be raised, depending on the tolerance of the specimen to irradiation damage. However, count rates will often be < 10 cps for the X-ray line of interest, especially in biological specimens. The operator has to accumulate sufficient counts at the peak position to differentiate from the counts at the background position. With such low counting rates this means relatively long counting times (> 100 sec) and care has to be taken to avoid errors due to contamination deposition, instrumental drift, beam current fluctuations, and extraneous electronic noise.

The important factor in determining a peak count is the ratio of peak-to-background counts. When this ratio is very low it eventually becomes impossible to distinguish the peak from the background. The minimum acceptable value of a peak intensity compared with its background depends on the actual counts accumulated. A useful criterion for determining the validity of a peak count over a background count where low counting rates are involved is as follows:

The height of the peak above zero is P, and the background is given by b. The statistical error in measuring $P - b$ is given by $(P + b)^{\frac{1}{2}}$. Thus for a significant value of P-b to exist it must be greater than $(P + b)^{\frac{1}{2}}$. The criterion is expressed as $P - b > n(P + b)^{\frac{1}{2}}$ where n is chosen depending on the reliability required. For example, if $n = 1$, a peak (P) count of 1000 and a background (b) count of 800 gives

$$\begin{aligned} 1000 - 800 &= 200 > (1800)^{\frac{1}{2}} \\ &\text{or } 200 > 42 = \text{correct.} \\ \text{If } P = 1000 \text{ and } b &= 960 \text{ however} \\ 1000 - 960 &= 40 > (1960)^{\frac{1}{2}} \\ &> 44 = \text{incorrect.} \end{aligned}$$

When working at low intensity levels repeated analyses should be performed to ascertain the reproducibility of the figures.

This criterion shows how the detection of very low concentration levels of elements depends on good peak-to-background ratios, and on high values of P. Thus analytical conditions must be adjusted (beam current, accelerating voltage, counting time, spectral line) to give the highest P values, while b is reduced to a minimum by the correct choice of specimen support (§ 4.1), specimen preparation procedures (§ 6.2.1a), etc. Some operators prefer to work with stricter tolerances and define the criterion for a valid peak detection as

$$(P - b) > 2(P + b)^{\frac{1}{2}}$$

Using this criterion it is possible to calculate the minimum detectable concentration of an element by the use of a standard of known concentrations. If C_m is the minimum detectable concentration, and P_m and b_m are the X-ray intensities given by this concentration, for a standard of concentration C_x, the X-ray intensities will be P_x and b_x. Thus

$$(P_m - b_m) = \frac{C_m}{C_x}(P_x - b_x).$$

Now $b_m \simeq b_x$ if the specimens are of similar composition, and for the minimum concentration $P_m \simeq b_m$.

The criterion defined above states that

$$(P_m - b_m) > n\,(P_m + b_m)^{\frac{1}{2}}$$

or

$$\begin{aligned} \frac{C_m}{C_x}(P_x - b_x) &> n\,(P_m + b_m)^{\frac{1}{2}} \\ &> n\,(2b_m)^{\frac{1}{2}} \\ &> n\,(2b_x)^{\frac{1}{2}} \end{aligned}$$

Thus

$$C_m = n \frac{C_x (2b_x)^{\frac{1}{2}}}{(P_x - b_x)}$$

where n is again chosen depending on the reliability required. For example, if C_x is 0.5% for an element x, and values obtained by analysis are $P_x = 1000$ and $b_x = 800$, the minimum detectable limit for that element under identical operating conditions would be

$$C_m = \frac{n \times 0.5 (2 \times 800)^{\frac{1}{2}}}{1000 - 800} \%$$

$$= 0.1n, \text{ or } 0.1\% \text{ if } n = 1.$$

References

Chandler, J. A. (1973), Recent developments in analytical electron microscopy, J. Microscopy *98*, 359.

Chandler, J. A. (1976), A method for preparing absolute standards for quantitative calibration and measurement of section thickness with X-ray microanalysis of biological ultrathin specimens in EMMA, J. Microscopy *106*, 291.

Cliff, G. and G. W. Lorimer (1975), The quantitative analysis of thin specimens, J. Microscopy *103*, 203.

Cobet, U. (1972), Quantitative electron beam microanalysis of biological thin tissue samples, 3rd Int. Conf. Med. Phys., Goteborg, Sweden, p. 324.

Cooke, C. J. and P. Duncumb (1966), Comparison of a non-dispersive method of X-ray microanalysis with conventional crystal spectrometry, in: Proc. 4th Int. Congr. X-ray optics and microanalysis, Orsay (Hermann, Paris), p. 467.

Duncumb, P. (1968), EMMA, combinaison d'un microscope électronique et d'une microsonde électronique, J. Microscopie *7*, 581.

Green, M. (1963), A Monte Carlo calculation of the spatial distribution of characteristic X-ray production in a solid target, Proc. phys. Soc. *82*, 204.

Hall, T. A. (1968), Some aspects of the microprobe analysis of biological specimens, in: Quantitative electron probe microanalysis, K. F. J. Heinrich, ed. (NBS Special pub. 298), p. 269.

Hall, T. A. (1971), The microprobe assay of chemical elements, in: Physical techniques in biochemical research, Vol. 1A, G. Oster, ed. (Academic Press, New York), p. 393.

Hall, T. A. (1974), The electron probe X-ray microanalysis of thin specimens, in: Advances in analysis of microstructural features by electron beam techniques (The Metals Society, London), p. 120.

Hall, T. A., H. C. Anderson and T. C. Appleton (1973), The use of thin specimens for X-ray microanalysis in biology, J. Microscopy *99*, 177.

Hall, T. A. and B. L. Gupta (1974), Beam induced loss of organic mass under electron microprobe conditions, J. Microscopy *100*, 177.

Hall, T. A. and P. D. Peters (1974), Quantitative analysis of thin sections and the choice of standards, in: Microprobe analysis as applied to cells and tissues, T. Hall, P. Echlin and R. Kaufmann, eds. (Academic Press, London & New York), p. 229.

Hall, T. A. and P. Werba (1969), The measurement of total mass per unit area and elemental weight fractions along line scans in thin specimens, 5th Int. Congr. X-ray optics and microanalysis, Tubingen, G. Mollenstadt and K. H. Gaukler, eds (Springer, Berlin), p. 93.
Hall, T. A. and P. Werba (1971), Quantitative microprobe analysis of thin specimens: continuum method, in: Proc. 25th Anniv. Meeting EMAG (Institute of Physics, London), p. 146.
Jacobs, M. H. and J. Baborovska (1972), Quantitative microanalysis of thin foils with a combined electron microscope microanalyser (EMMA-3). Proc. 5th Eur. Reg. Conf. Electron Microscopy, Manchester, p. 136.
Marshall, D. J. and T. A. Hall (1968), Electron probe X-ray microanalysis of thin films, Br. J. appl. Phys. (D), *1*, 1651.
Philibert, J., J. Rivory, D. Bryckaert and R. Tixier (1970), Electron probe microanalysis on electron microscope thin foils using thin standards, Met. Phys. *479*, 68.
Philibert, J. and R. Tixier (1968), Some problems with quantitative electron probe microanalysis, in: Quantitative electron probe microanalysis, K. F. J. Heinrich, ed. (NBS Special pub. 298, Washington, D.C.), p. 13.
Russ, J. C. (1974), The direct element ratio model for quantitative analysis of thin sections, in: Microprobe analysis as applied to cells and tissues, T. Hall, P. Echlin and R. Kaufmann, eds (Academic Press, London), p. 269.
Spurr, A. R. (1975), Choice and preparation of standards for X-ray microanalysis of biological materials with special reference to macrocyclic polyether complexes. In: Biological microanalysis, P. Galle and P. Echlin, eds (Soc. Française de Mic. Electronique, Paris), p. 238.
Sweatman, T. R. and J. V. P. Long (1969), Quantitative electron probe microanalysis of rock forming minerals, J. Petrol. *10*, 332.
Sweeney, W. E., R. E. Seebold and L. S. Birks (1960), Electron probe measurements of evaporated metal films, J. appl. Phys. *31*, 1061.
Warner, R. R. and J. R. Coleman (1974), Quantitative analysis of biological material using computer correction of X-ray intensities, in: Microprobe analysis as applied to cells and tissues, T. Hall, P. Echlin and R. Kaufmann, eds (Academic Press, London & New York), p. 249.
Yakowitz, H. (1975), Computational schemes for quantitative X-ray analysis: on-line analysis with small computers, in: Practical scanning electron microscopy. Electron and ion microprobe analysis, J. I. Goldstein and H. Yakowitz, eds (Plenum Press, New York).

Further reading

Quantitative X-ray microanalysis

Beaman, D. R. and J. R. Isasi (1972), Electron beam microanalysis. ASTM special publication 506, (ASTM, Pennsylvania).
Birks, L. S. (1969), X-ray spectrochemical analysis, 2nd edn. (John Wiley, New York).
Duncumb, P. and P. K. Shields (1963), The present state of quantitative X-ray microanalysis, part 1: physical basis. Brit. J. appl. Phys. *14*, 617.
Goldstein, J. I. and H. Yakowitz eds (1975), Practical scanning electron microscopy. Electron and ion microprobe analysis. (Plenum Press, New York).
Hall, T. A. (1971), The microprobe assay of chemical elements, in: Physical techniques in biological research, 2nd Edn, Ed. G. Oster, (Academic Press, New York), p. 393.
Heinrich, K. F. J. (1968), Quantitative electron probe microanalysis, NBS special publication 298, (NBS, Washington, D.C.).

Martin, P. M. and D. M. Poole (1971), Electron probe microanalysis: the relation between intensity ratio and concentration, Metallurgical Rev. *150*, 19.
McKinlay, T. D., K. F. J. Heinrich and D. B. Wittry (1966). The electron microprobe (John Wiley, New York).
Philibert, J. (1970), Electron probe microanalysis, in: Techniques of metals research, Vol. III, part 2, Modern analytical techniques for metals and alloys, R. Bunshah, ed. (Wiley Interscience, New York).
Poole, D. M. and P. M. Martin (1969), Electron probe microanalysis: Instrumental and experimental aspects, Metallurgical Rev. *133*, 61.

Chapter 7

Applications of X-ray microanalysis in the electron microscope

To illustrate many of the practical considerations previously discussed, analyses of some metallurgical, mineralogical and biological specimens will now be described in detail.

7.1 *Metallurgical applications*

7.1.1 *Analysis of a metal foil in EMMA*

Figure 7.1 shows a transmission electron micrograph of a region near a grain boundary in a metal foil of an Al/Zn alloy, heat treated for 1 hr at 500 °C and then for 4 hr at 240 °C.

The electron probe was focused into the regions shown by the circles and the crystal spectrometers set to detect Al and Zn simultaneously. The Zn/Al count rate ratio at each point analysed is shown in Fig. 7.2. This ratio was corrected to provide elemental ratios as described in § 6.2.2. For the analysis a beam current of 20 nA and a counting time of 100 seconds were employed at each point, with an accelerating voltage of 100 kV.

The curve (Fig. 7.2) shows that zinc is depleted in a region on either side of the grain boundary, but that grain boundary precipitates contain high concentrations of zinc.

7.1.2 *Analysis of a metal extraction replica in the TEM.*

Figure 7.3 shows an extraction replica of an oxide layer formed on a Fe/10 % Cr alloy surface after breakaway oxidation at 500 °C (Fursey 1971). The oxide was mounted on a copper EM grid for analysis in transmission. With

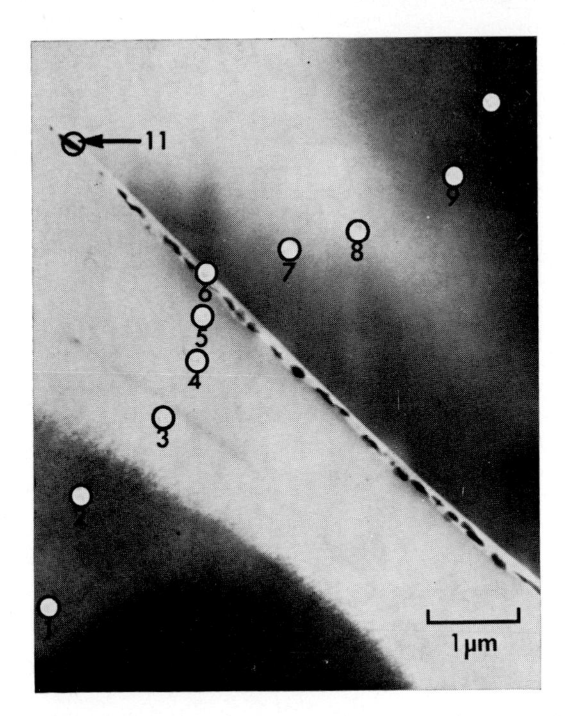

Fig. 7.1. Analysis across a grain boundary in an Al/Zn alloy thin metal foil. The circles represent the positions of the probe. (Courtesy of Dr. M. Jacobs, T.I. Laboratories, Cambridge.)

one crystal spectrometer set to detect CrK_α and the other FeK_α, each part of the specimen was analysed by focusing the electron beam onto the relevant regions. Analysis conditions used were: accelerating voltage, 100 kV; beam current, 0.02 μA; and counting time 100 sec. Both CrK_α and FeK_α were measured on LiF crystals in crystal spectrometers.

Large oxide nodules with protruding whiskers were found to contain Fe and Cr in a ratio of 6:1, whereas in the smaller nodules without whiskers the Fe/Cr ratio was 16:1.

One possible source of error in such an analysis using twin crystal spectrometers to determine the elemental ratio of Fe/Cr, is that one of the signals passing towards one of the crystal spectrometers may be shadowed by a grid bar (§ 5.3.2). To check this possibility each spectrometer should be tuned to detect each of the elements in turn.

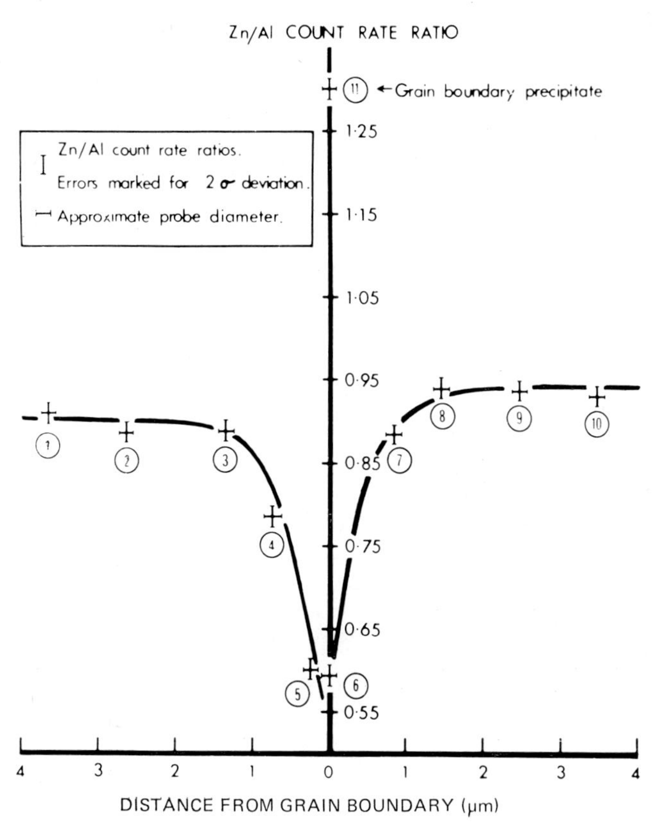

Fig. 7.2. X-ray count rate ratios for Zn/Al in the regions numbered in Fig. 7.1. (Courtesy of Dr M. Jacobs, T.I. Laboratories, Cambridge.)

7.1.3 Analysis of a metal oxide extraction replica in the SEM

Figure 7.4 shows SEM X-ray images of an oxide film of Fe/Cr similar to the one just described. In Fig. 7.4a the X-ray detector was tuned to detect X-rays of the FeK_α characteristic line, and the X-ray intensity output was synchronised with the SEM raster so that bright dots occurred where there was a high content of Fe. In Fig. 7.4b, the detector was tuned to receive the CrK_α characteristic line and the distribution of Cr within the oxide film was found to be different to that of Fe (Fursey 1971). The time of analysis required to accumulate the X-ray distribution picture was approximately 100 sec in this sample where fairly high concentrations ($> 10\%$) of each element were present. A typical analysis was performed at an accelerating voltage of 30 kV, a beam current of 0.1 μA and a focusing aperture of 400 μm.

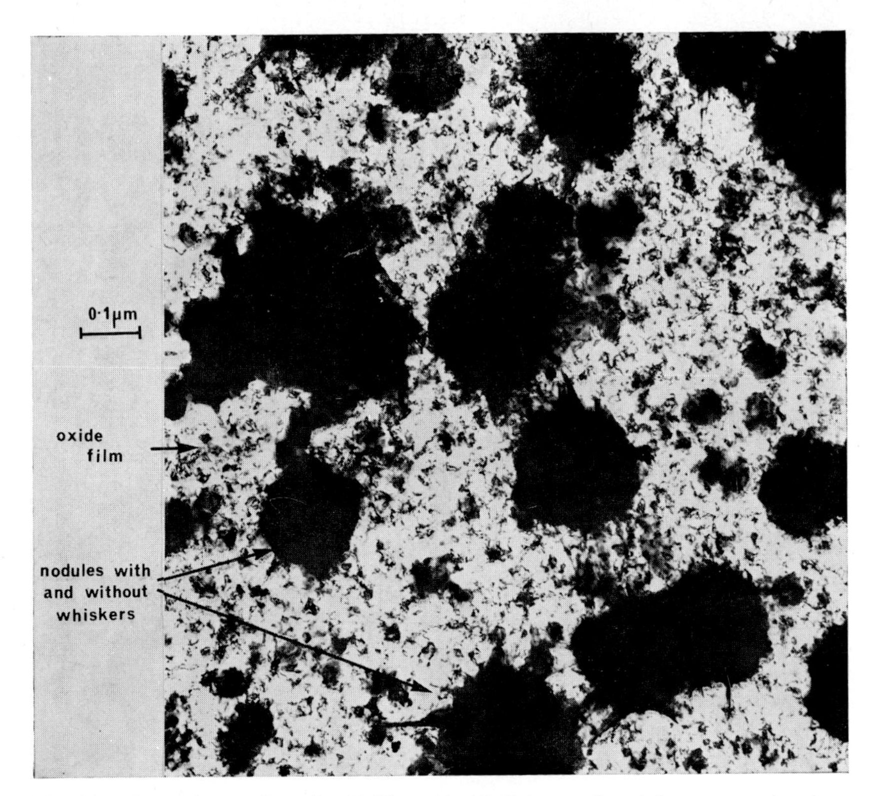

Fig. 7.3. Extraction replica of a Fe/Cr oxide. Both types of nodules were analysed as well as the background.

7.2 Mineralogical applications

7.2.1 Analysis of a powder dispersion in EMMA

Figure 7.5 shows a dispersion of particles of montmorillonite, a clay mineral prepared by drying a suspension in ethanol onto a carbon-coated grid (§ 4.3.2). These particles are sometimes less than 10 nm thick and are barely visible in the electron microscope. To protect them against electron beam damage they are coated with a thin film of carbon evaporated *in vacuo.*

Individual particles were selected for analysis and the electron probe focused to surround each one in turn. The beam current was carefully controlled (in this experiment less than 10 nA at 80 kV) to avoid specimen damage. Time of analysis was 100 sec or more depending on the count rate. With such thin particles longer count rates may be necessary and then care

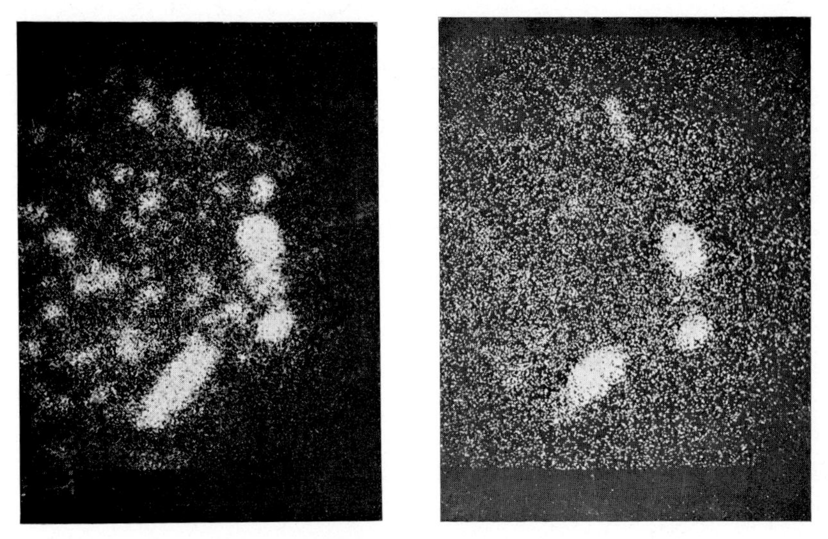

Fig. 7.4. (a) X-ray image of FeK_α emission from a Fe/Cr oxide extraction replica. (b) X-ray image of CrK_α emission from the same extraction replica.

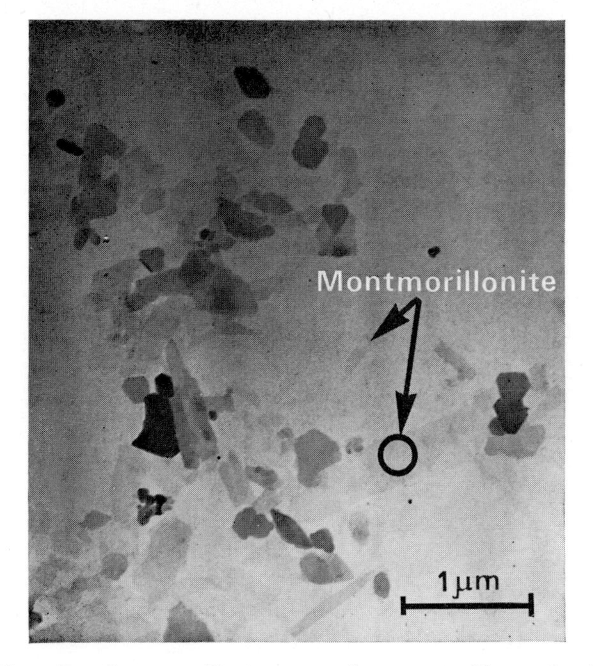

Fig. 7.5. Dispersion of montmorillonite on a carbon support film, analysed in EMMA.

must be taken to ensure that elemental loss does not occur (§ 5.1.2).

For determining the whole elemental spectrum an SSD was employed, while individual elemental peaks were detected with crystal spectrometers when high peak-to-background ratios were required (see § 3.3.2). The ratio method of quantitation was used (§ 6.2.2), and appropriate standards (§ 4.5.2) allowed accurate determination of elemental ratios in individual particles, and absolute quantitation of separate elements.

7.3 *Biological applications*

7.3.1 *Analysis of a thin biological section in the TEM/EMMA*

Figure 7.6 is a micrograph of a section of rat lateral prostate epithelium in which the subcellular distribution of zinc was investigated using the pyro-antimonate technique (§ 4.4.2b) to ensure the retention of the cation during tissue fixation. Precipitates resulting from the reaction were seen within the epithelium. Sections 90 nm thick were mounted on aluminium grids and then coated with a thin (10 nm) layer of carbon for protection. Since very low levels ($< 0.1\,\%$) of zinc were to be detected, crystal spectrometers were used in preference to a solid state detector because of the better peak-to-background values obtained. The sections were mounted in a special holder made from beryllium (Chandler 1973) to reduce X-ray background and eliminate copper X-ray lines resulting from stray electrons (§ 6.2.1a).

The electron beam (0.1 μA) was focused onto subcellular organelles (diameter 0.1 μm) and the ZnK_α line emission counted for 40 sec. The spectrometer was then tuned away from the peak wavelength and a background signal recorded for 40 sec (§ 5.3.3c). Each organelle or subcellular region was analysed in the same way.

An advantage of using crystal spectrometers in this application is made apparent by examining the spectrum obtained at the region of the ZnK_α wavelength (Fig. 7.7). The preparation of the specimen involved the use of OsO_4 during fixation and thus Os was incorporated into the tissue and was present in the cells. The OsL_α line is very close to the ZnK_α line and was much greater in intensity due to the large amount of Os present. An SSD would not enable these two peaks to be separated, especially with such a large intensity difference, and deconvolution techniques (§ 5.3.4g) would not be able to manipulate the relative intensities. The crystal spectrometers have sufficiently good energy resolution to permit this separation.

Analysis of several cells in the prostate epithelium revealed that zinc

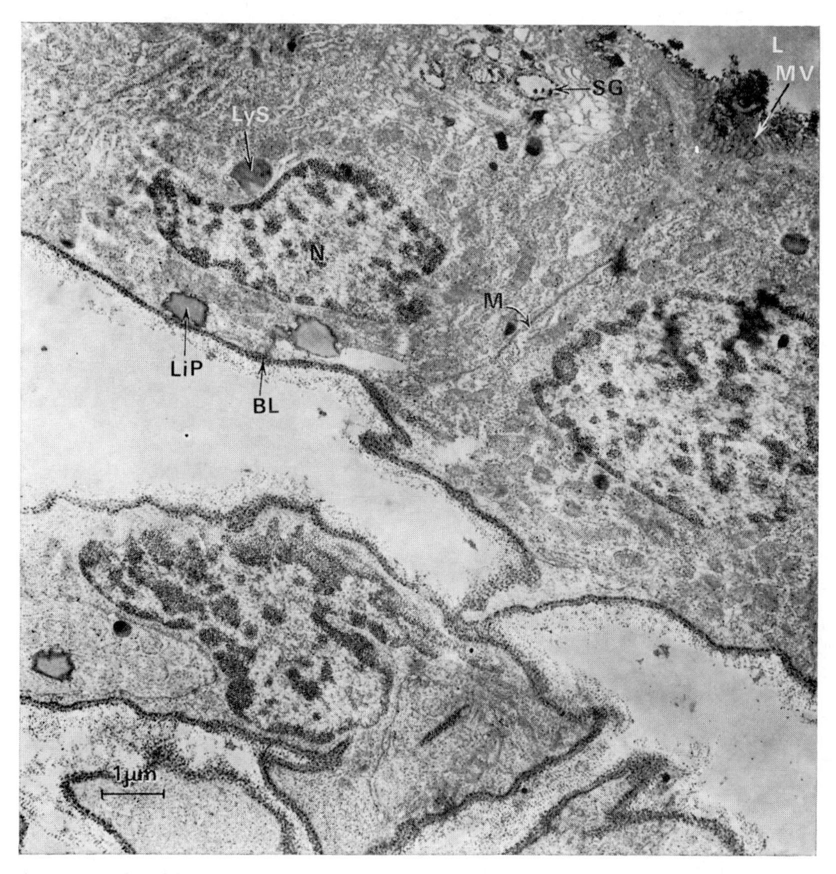

Fig. 7.6. Ultrathin section of rat prostate fixed with potassium pyroantimonate and unstained for analysis in EMMA. The antimony precipitates are associated with zinc and calcium. LyS = lysosome, SG = secretory granule, N = nucleus, LiP = lipid, BL = basal lamina, M = membrane, MV = microvilli, L = lumen.

is most highly concentrated in the nucleolus, chromatin, lysosomes and secretory granules (Chandler et al. 1977). Calculation of elemental concentrations in subcellular regions was performed by comparison with the emission from a resin standard containing a fixed concentration of potassium (§ 4.5.2b) and corrected for relative detection efficiency (§ 6.3). The concentrations found in various regions of the tissue are shown in Fig. 7.8.

In any biological analysis such as this, results are only meaningful if they have statistical significance. In biological tissue there may be a very wide range of physiological states existing between cells and organelles and it is therefore essential to perform as many analyses as will truly represent

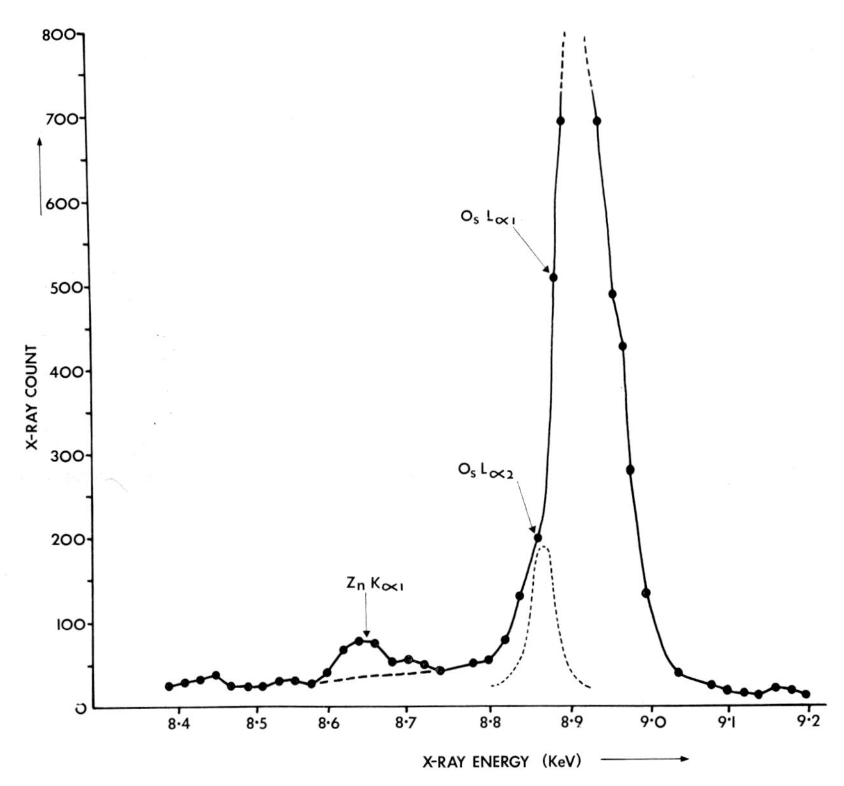

Fig. 7.7. The peak in the crystal spectrometer spectrum for ZnK_{α} is adjacent to the peak for OsL_{α}. (Courtesy of John Wiley and Sons, Inc., New York)

the whole tissue. This may mean in practice the analysis of over 100 organelles of each type, covering many cells in many randomly chosen areas of the tissue. It is usually insufficient to quote analyses from just a few subcellular regions when trying to attach biological significance to the results.

7.3.2 Analysis of an ultrathin frozen section in the TEM

Figure 7.9 is a micrograph of an ultrathin section of unfixed frozen frog skeletal muscle, prepared by freezing with liquid nitrogen (Sjöström and Thornell 1975). Dry sections were cut frozen at −100 °C on an ultramicrotome, collected on carbon-coated copper grids and freeze-dried in dry nitrogen gas.

Analysis of selected regions was performed in the TEM with a focused electron probe, an accelerating voltage of 80 kV and a counting time of

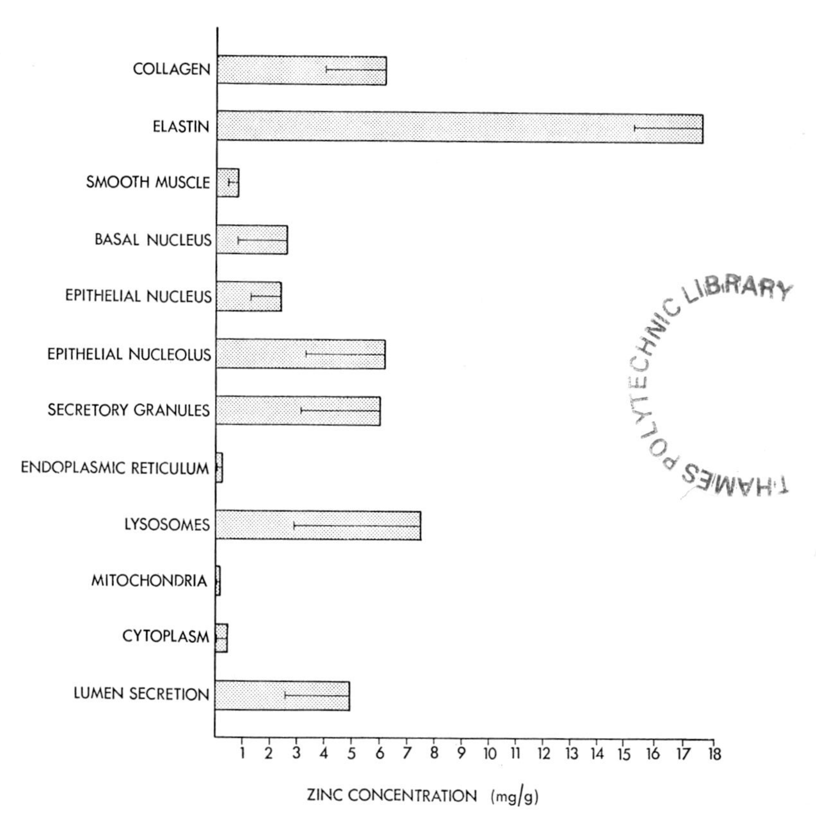

Fig. 7.8. Subcellular concentrations of zinc in rat lateral prostate (see Fig. 7.7).

200 sec per region. The X-ray spectrum obtained from the multichannel analyser associated with the SSD is shown in Fig. 7.10 and indicates the presence of Mg, Si, P, S, Cl, K and Ca in the A band of the muscle. In this tissue similar analyses with fixed tissue fail to demonstrate the presence of Mg, P, K or Ca. Thus frozen sections are required to retain certain elements, notably electrolytes, that may be extracted during fixation or other procedures.

7.3.3 *Analysis of air-dried sperm cells in the TEM/EMMA*

Specimens of human sperm cells (Fig. 7.11) from 20 donors were prepared for analysis by air-drying drops of semen onto nylon electron microscope grids covered with carbon support films. The cells were then coated with carbon by evaporation and analysed in an EMMA. Previous investigations (Chandler and Battersby 1976) had determined that air-drying caused no

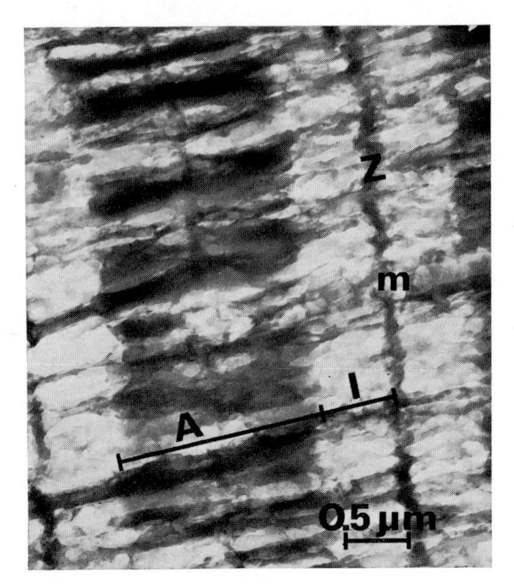

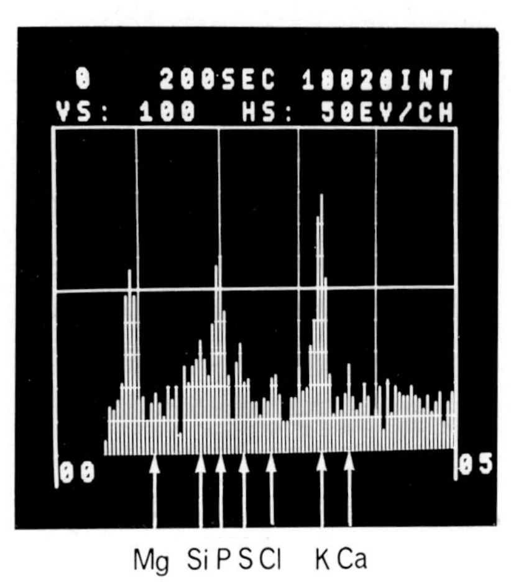

Fig. 7.9. Ultrathin frozen and freeze-dried section of frog skeletal muscle analysed in the TEM.

Fig. 7.10. SSD spectrum of elements analysed in the A band of the muscle section shown in Fig. 7.9.

difference in elemental distributions as compared with freezing and freeze-drying.

The cells were analysed in the acrosome, nucleus and midpiece regions with a static probe. The accelerating voltage was 80 kV and a beam current of 0.05 μA was used for 40 sec in each subcellular region. The conditions of analysis were found not to affect specimen composition (Battersby 1976; Chandler and Battersby 1976).

The solid state detector was set to measure, by integration, the line intensities of Na, Mg, P, S, Cl, K, Ca and Zn and to monitor the white radiation simultaneously. Using resin standards (§ 4.5.2b) the absolute dry mass fraction of each element was calculated. The elemental spectrum is shown in Fig. 5.16 and the elemental concentrations, based on analysis of 20 cells for 20 donors, is shown in Fig. 7.12.

The great advantage of the SSD in this type of application lies in providing simultaneous measurement of all 8 elements together, as well as of specimen mass thickness. When linked to a computer, the analytical system can provide elemental concentrations immediately.

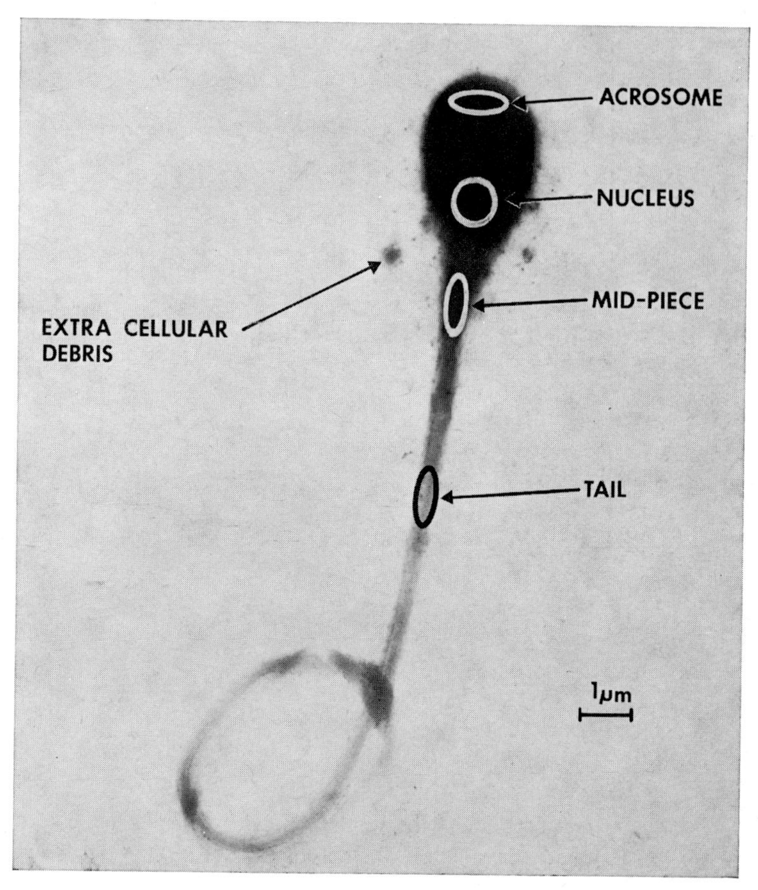

Fig. 7.11. Air-dried human sperm cells analysed in EMMA. The subcellular regions of analysis are indicated by the circles. (Courtesy of John Wiley and Sons, Inc., New York.)

7.3.4 Analysis of a frozen-hydrated biological section in the STEM

Figure 7.13 shows a frozen-hydrated section of a *Rhodnius* malphigian tubule cut at a thickness of 1 μm and examined in the scanning microanalyser with transmission electron detection. The sections were cut from tissue that had been frozen and maintained in the hydrated state according to the method of Saubermann and Echlin (1975).The tissues were rapidly quench frozen in Isceon 13 (mono-chlorotrifluoromethane, ISC Chemicals Ltd.) cooled to its melting point (−181 °C) with liquid nitrogen, and then transferred to liquid

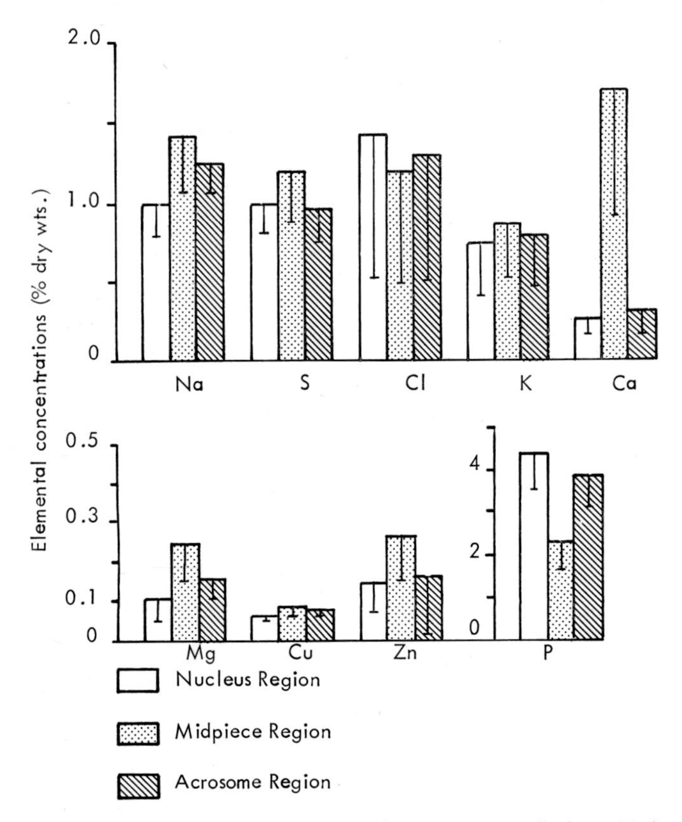

Fig. 7.12. Elemental composition of air dried human sperm cells from 20 donors. The bars represent 1 s.d. of the mean.

nitrogen. Blocks were trimmed and sections cut on a cryostat at $-85\,^{\circ}$C (SLEE, see Appendix) to a thickness of 1–2 μm. They were mounted on a nylon film and coated with a thin layer of aluminium, after which they were inserted into an SEM via a transfer specimen holder. The sections were mounted on a cold stage in the SEM and examined at low temperatures by STEM techniques. At no stage in the process were the sections allowed to thaw or to dry. A mass spectrometer was used to monitor losses of water and organic material from the specimen. The authors describe in some detail the procedures necessary to maintain the frozen hydrated state of the specimen and suggest a number of criteria for testing the maintenance of the specimen's hydration. It was estimated that only 10% of the water was lost, sufficient however to provide image contrast. The analysis was performed

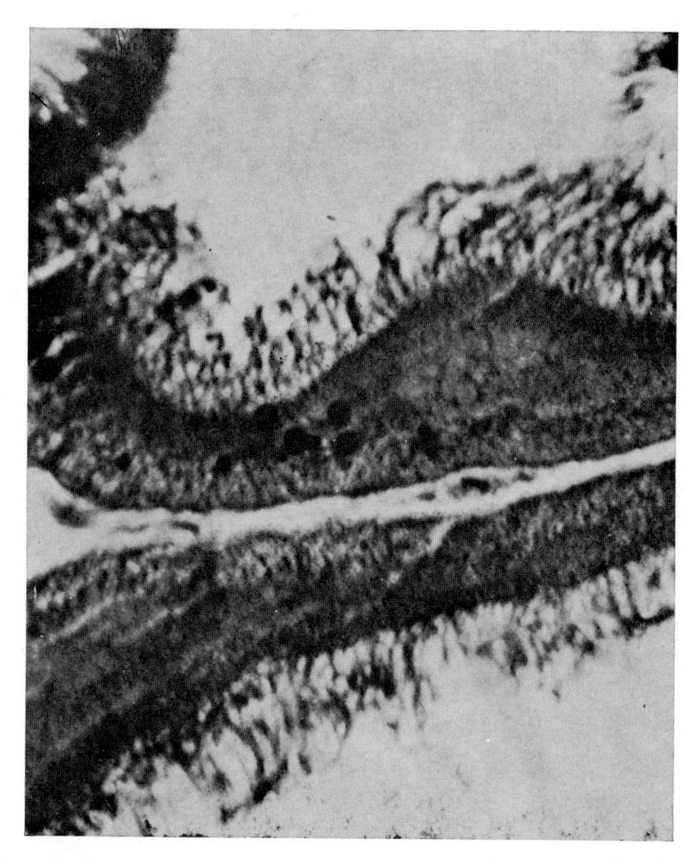

Fig. 7.13. Scanning transmission image of a frozen hydrated section (1 μm) of malphigian tubule, maintained at −150 °C. The brush border and basement membrane are clearly seen. (Courtesy of Dr. B. L. Gupta, Zoology Department, University of Cambridge.)

with the specimen maintained at −150 °C. The image shows brush border and basement membrane clearly.

Analysis was performed with a beam current of 0.5–5 nA, at 50 kV, and with a static beam probe of 100 nm. Distributions of electrolytes (Na, K, Cl) were clearly distinguished in subcellular regions of the specimen.

This type of analysis illustrates the problems of correlating ultrastructural detail with subcellular analysis when using cryo-preparation techniques. At the present time, however, it appears to be a most promising method for the analysis of electrolytes in tissue.

References

Battersby, S. (1976), Electron microscope microanalysis of trace metals in human sperm, M.Sc. Thesis, University of Wales.

Chandler, J. A. (1973), Recent developments in analytical electron microscopy, J. Microscopy *98*, 359.

Chandler, J. A. and S. Battersby (1976), X-ray microanalysis of ultrathin frozen and freeze dried sections of human sperm cells, J. Microscopy *107*, 55.

Chandler, J. A., B. G. Timms and M. S. Morton (1977), Subcellular distribution of zinc in rat prostate studied by X-ray microanalysis, I. Normal prostate, Histochem. J. *9*, 103.

Fursey, A. (1971), Metal oxide structure investigated using a variety of light- and electron-microscopical techniques. J. Microscopy *94*, 167.

Saubermann, A. J. and P. Echlin (1975), The preparation, examination and analysis of frozen hydrated tissue sections by scanning transmission electron microscopy and X-ray microanalysis, J. Microscopy *105*, 155.

Sjöström, M. and L. E. Thornell (1975), Preparing sections of skeletal muscle for transmission electron analytical microscopy (TEAM) of diffusible elements, J. Microscopy *103*, 101.

Further reading

The following is a list of publications containing a number of useful references to analyses of biological and non-biological specimens in both the SEM and TEM.

(a) *Biological*

Hall, T. A. (1971), The microprobe assay of chemical elements, in: Physical techniques in biochemical research, Vol. 1A, 2nd edn. G. Oster, ed. (Academic Press, New York and London), p. 158.

Hall, T., P. Echlin and R. Kaufmann, (eds) (1974), Microprobe analysis as applied to cells and tissues (Academic Press, London and New York).

Hall, T. A., H. D. E. Rochert and R. L. De C. H. Saunders (1972), X-ray microscopy in clinical and experimental medicine (Charles Thomas, Illinois).

Echlin, P. and P. Galle, eds (1975), Biological microanalysis (Societé Française de Microscopie Électronique, Paris).

Mollenstadt, G. and K. H. Gaukler (1969), X-ray optics and microanalysis (Springer, Berlin).

Russ, J. C. and B. J. Panessa (1972), Thin section microanalysis (Edax Laboratories Ltd, North Carolina, U.S.A.).

Various proceedings of the SEM Symposia, O. M. Johari, ed. (ITT Research Institute, Chicago, Ill.).

Various proceedings of the Electron Probe Analysis Society of America.

(b) *Materials science*

Beaman, D. R. and J. A. Isasi (1972), Electron beam microanalysis, ASTM STP 506 (ASTM, Philadelphia, U.S.A.).

McKinley, T. D., U. F. J. Heinrich and D. B. Wittry (1966), The electron microprobe (Wiley, New York).

Mollenstadt, G. and K. H. Gaukler (1969), X-ray optics and microanalysis (Springer, Berlin).

Goldstein, J. I. and H. Yakowitz (1975), Practical scanning electron microscopy. Electron and ion microprobe analysis (Plenum Press, New York).

Various proceedings of The Electron Probe Analysis Society of America.

Various proceedings of the SEM Symposia, O. M. Johari, ed. (ITT Research Institute, Chicago, Ill.).

Lorimer, G. W., N. A. Razik and G. Cliff (1973), The use of the analytical electron microscope EMMA-4 to study the solute distribution in thin foils – some applications to metals and minerals, J. Microscopy *99*, 153.

Lorimer, G. W. and G. Cliff (1976), Analytical electron microscopy of minerals, in: Electron microscopy in mineralogy, (Springer-Verlag, Berlin), p. 506.

Lorimer, G. W., M. J. Nasir, R. B. Nicholson, K. Nuttall, D. E. Ward and J. R. Webb (1972), The use of an analytical electron microscope (EMMA-4) to investigate solute concentrations in thin metal foils, In: Electron microscopy and the structure of materials, G. Thomas, ed. (University of California Press, Berkeley), p. 222.

Chapter 8

Some common problems

8.1 *What equipment to buy (or use)*

Firstly the investigator must decide the nature of the problem to be solved. Usually a piece of equipment is bought on a group basis, a joint application is made for funds, and a wide range of analytical problems will be tackled, although usually within the same discipline (i.e. biology or materials science).

The decision must be made as to whether electron microscopy or X-ray microanalysis is the most important technique since a compromise is sometimes necessary in the design of an instrument. If relatively high resolution electron microscopy is the main requisite, so that the analysis can be correlated with precise visualization of ultrastructure, then a transmission electron microscope will be required. If the finer details of the ultrastructure are not so important and the interests of the laboratory are in accurate X-ray analysis, then a scanning microscope may be the choice. Thus, the first consideration is that of image resolution.

If a thick sample (say up to 2 μm) is to be examined, and the SEM is the optical system chosen, there then lies a choice between a crystal spectrometer or an SSD for X-ray analysis. The parameters discussed in § 3.3 and listed in Table 3.3 will help determine the choice. It is extremely difficult to generalise because each specimen and each type of application requires its own special treatment. However, some guidelines can be given for the various categories of application mentioned below.

8.1.1. *Detection of trace elements*

Trace elements (< 1 %) require good peak-to-background ratios for detection. This may suggest that a crystal spectrometer is the best choice, especially if a single or limited number of elements are being detected repeatedly. However, if qualitative elemental spectra are to be determined from a specimen then the SSD is easily the most convenient in that it provides a complete energy spectrum. The detection sensitivity of the system depends on the geometry of the design. In the SEM there is ample room to place a detector close to the specimen for maximum solid angle of collection. With the crystal spectrometer the focusing requirements of the crystals limit the solid angle of collection but detection efficiency for single X-ray lines is often comparable and even superior to the SSD (Chandler 1973). With biological specimens there is a need to limit the specimen current to avoid damage. Thus collection efficiency is an important factor.

By far the majority of SEMs have SSD attachments, mostly due to the greater ease of fitting and using them, but the characteristics of the specimen energy spectrum should be carefully examined before deciding on the SSD. Energy resolution is an important factor. The ability to separate peaks is better in the crystal spectrometer than in the SSD and if there are overlapping peaks in the spectrum (as in the example discussed in § 7.3.1) crystal spectrometers may be essential.

In the TEM and EMMA the structure of the specimen stage is such that it is difficult to prevent extraneous X-rays from entering the detection system. The SSD collects far more of these X-rays than the crystal spectrometer because the X-rays are not focused. The peak-to-background ratio may be seriously affected if the SSD is not adequately shielded and spurious X-ray lines can more easily interfere with the spectrum.

8.1.2 Detection of major elements

When high concentrations ($>5\%$) of elements are present in the specimen, peak-to-background ratios are usually high and the SSD may be more convenient. However, the same problem of overlapping X-ray lines may occur and the investigator must determine whether the SSD can resolve the peaks. The greatest advantages of the SSD in quantitative work with higher concentrations of elements are the simultaneous display of peak and background in the energy spectrum, and the ability to compute concentrations from the integrals of the peaks. This avoids the possible source of error in crystal spectrometers of irreproducibility of the setting of the peak wavelength positions.

As an example, the analysis of human sperm cells shown in § 7.3.3 is far more readily done on a routine basis with the SSD since many elements can be analysed simultaneously in a matter of seconds. However, in the analysis illustrated in § 7.3.1 the SSD could not separate the overlapping lines of Zn and Os and the crystal spectrometer was essential.

In general, therefore, it is a great advantage to have both crystal spectrometers and SSD systems fitted simultaneously. Unfortunately this is not always possible and most existing SEMs and TEMs are more readily fitted with SSD systems alone. Such systems are extremely useful but the investigator should pay much attention to certain factors which may affect the sensitivity of the analysis as summarised below (§ 8.4).

8.1.3 Choosing total or partial systems

A further consideration is, of course, the cost. Many existing SEMs and TEMs can be fitted subsequently with X-ray attachments, or the complete system can be initially purchased. The various combinations are:

A. SSD and MCA (minimum requirement);
 + chart recorder (for recording spectra)
 + computer (for processing data)
 + magnetic tape (for storing data)
 + teletype (for output of data onto paper)

B. Spectrometer (one or two) and ratemeter, PHA and scaler (minimum requirement);
 + oscilloscope (for visualising pulses)
 + scan control (for scanning automatically through the spectrum)
 + chart recorder (for recording spectrum)
 + servo controls (for pre-selecting wavelengths).

In both A and B the cost increases as systems are added and the investigator must decide on the requirements beforehand. This can only be done by fully appreciating the value of each item as described in previous chapters.

8.2 Technical expertise required

There is no use pretending X-ray microanalysis is an easy procedure. Those who have been practising electron microscopists for many years will know that many hidden snags and pitfalls lie in wait to trap the unwary. However, an investigator with a good working knowledge of electron microscopy, who knows not only how to prepare specimens and take micrographs but also *understands* how the electron microscope works, will have little difficulty in understanding the rudiments of X-ray microanalysis and in becoming familiar with the various procedures described here. Certainly a person with a background in physics, rather than biology, has a great advantage in appreciating the mode of operation of the instrument. It is a technique which requires a good deal of experience to perfect, plus an ability to sometimes grapple with mathematical formulae. Possibly the most difficult part of any operator's training is in the handling of mathematical procedures for quantitative analysis.

Many instruments today have been designed so that results can be obtained by almost anyone by the push of a button. Many simple applications may fall into the category of 'button pushing for data'. However, detailed analysis of specimens for metallurgical, mineralogical and biological applications require more than this and the microscopist to whom the task is given should be familiar with the nature of the specimen; the effect of preparation procedures on the specimen; the effect of the electron beam on the specimen; the effect of the operating conditions on the X-ray data; the physical parameters involved in obtaining maximum quantitative information from the X-ray detector and the mathematics involved in interpreting the X-ray information; and last, but not least, should have the ability to correlate this information with the structural information contained in the image of the specimen. Such people are rare and more frequently a group of people will share the necessary experience. If such a group is not available, then it is probably worthwhile employing an assistant who can learn to operate and look after the instrument efficiently and who can communicate well with the people who have the scientific problems to solve.

Companies supplying instruments often provide training courses for users and it is strongly recommended that use is made of these after a few months of operation of an instrument.

8.3 Maintenance of equipment

As with a conventional electron microscope it is strongly recommended that a maintenance contract is arranged between the user and the company supplying the equipment. This is particularly necessary for the maintenance or repair of the nucleonics, much of which may be modular in design and is simply replaced by plug-in units when faulty.

However, competent technical laboratory staff should be able to maintain the electron-optical part of the instrument. This means giving regular and routine attention to the following:

(i) Cleanliness of those parts of the microscope under vacuum;

(ii) Replacement of burnt-out filaments;
(iii) Alignment of the microscope column;
(iv) Replacement of X-ray windows (when fitted);
(v) Replacement and cleaning of apertures;
(vi) Alignment of crystal spectrometers (when fitted);
(vii) Calibration of MCA energy scale and range.

The room in which the microscope is installed must be dust-free, vibration-free and have a stable temperature. It is recommended that a closed-circuit water system is used. Other important requirements are described by Alderson (1975).

8.4 *Factors affecting sensitivity*

Frequently an operator may be frustrated by an apparent lack of sensitivity in the instrument in an analysis of a specimen containing very low concentrations of elements. The following check list of possible sources of insensitivity may help to diagnose the problems and help achieve better performance:

(i) Choice of accelerating voltage (§ 5.2.1)
(ii) Choice of beam current (§ 5.2.2)
(iii) Choice of spectral line (§ 5.2.6a)
(iv) Choice of crystal (§ 5.2.6b)
(v) Choice of pressure (§ 5.2.6c + § 5.2.7c)
(vi) Thickness of X-ray window (§ 5.2.6d)
(vii) Length of analysis (§ 5.2.4)
(viii) Choice of grid material and mesh size (§ 5.1.4)
(ix) Loss of elements in electron beam (§ 5.1.2 + § 6.4.1)
(x) Contamination (§ 5.1.3 + § 6.4.1)
(xi) Diameter of electron probe (§ 5.3.2)
(xii) Removal of objective aperture (§ 5.2.3)
(xiii) Removal of anti-contaminator blade (§ 5.2.3)
(xiv) High background from stains in section (§ 6.2.1a)
(xv) Section thickness (§ 3.1.4)
(xvi) Spurious contributions to X-ray background (§ 6.2.1a)
(xvii) Adjustment of PHA (§ 5.2.6f)
(xviii) Integration of peaks versus peak height measurement (§ 5.3.4d)
(xix) Setting of crystal spectrometer (§ 5.3.3b)

TABLE 8.1
The detection of elements in thin specimens

Element	*Line*	*Fluorescent yield*	*Ease of detection*	*Stability during preparation (biol)*	*Stability during irradiation*	*Confusing lines*	*Other comments/ section references*
$Z < 11$	K_α	<0.01	Difficult	Unknown	Variable	—	Low kV required. Not usually detected in thin specimens
Sodium 11	K_α	0.02	Moderate	Poor[a]	Poor	CuL_α	[a] Frozen sections required (biol) § 4.4.2a
Magnesium 12	K_α	0.03	Moderate	Poor[a]	Poor	NaK_β GeL_α	[a] Frozen sections required (biol) § 4.4.2a
Aluminium 13	K_α	0.04	Medium	[a]	Good		[a] Not often found in biological systems
Silicon 14	K_α	0.055	Good[a]	Good[a]	Good	OsM_α	[a] Contamination from some resins/vacuum oil
Phosphorus 15	K_α	0.07	Good	Variable[a]	Good	OsM_α	[a] Extraction of phospholipids by solvents
Sulphur 16	K_α	0.09	Good	Good[a]	Good	PK_β, MoL_α, PbM_α	[a] Contamination from resins/vacuum oil
Chlorine 17	K_α	0.105	Good	Poor[a]	Poor	SK_β	[a] Frozen sections required
Potassium 19	K_α	0.14	High	Poor[a]	Variable	UM_α	[a] Frozen sections required
Calcium 20	K_α	0.19	High	Variable[a]	Good	SbL_α SnL_β, KK_β	[a] Histochemical/freezing techniques used, § 4.4.2
Titanium 22	K_α	0.22	High	[a]	Good	BaL_α, ScK_β	[a] Not often found in biological samples. Ti in nylon grids. Spurious signals from microscope pole pieces etc.
Vanadium 23	K_α	0.24	High		Good	TiK_β	
Chromium 24	K_α	0.26	High		Good	VK_β	
Manganese 25	K_α	0.28	High		Good	CrK_β	
Iron 26	K_α	0.32	High	Good	Good	MnK_β	Spurious signals from microscope pole pieces, etc.

Cobalt 27	K_α	0.35	High	Unknown	Good	FeK_β	Histochemical final reaction product
Nickel 28	K_α	0.37	High	Unknown[a]	Good	[b]	[a] Not often found in biological specimens [b] Spurious signals from microscope pole pieces, etc.
Copper 29	K_α	0.41	High	Good	Good	[a]	[a] Spurious signals from grids, pole pieces, etc.
Zinc 30	K_α	0.43	High	Good[a]	Variable[b]	OsL_α CuK_β	[a] Histochemical methods used, § 4.4.2b [b] Sometimes volatile
Arsenic 33	K_α L_α	0.53	High	Unknown[a]			[a] Present in cacodylate buffer
Bromine 35	K_α L_α	0.60	[a]		Very volatile	HgL_β	[a] Extremely hard to analyse due to radiation losses
Tin 50	K_α L_α	0.86 0.11	Good	[a]	Good	[b]	[a] Not often found in biological samples [b] Present in copper grids
Antimony 51	K_α L_α	0.87 0.12	Good	[a]		CaK_α	[a] Used as a cytochemical stain § 4.4.2b
Iodine 53	K_α L_α	0.88 0.13	[a]	[b]	Very volatile		[a] Very difficult due to volatility during irradiation [b] Used as a biochemical label
Mercury 80	L_α M_α	0.37	[a]	Variable[b]	Very volatile[b]		[a] Very difficult due to volatility during irradiation [b] Depends on binding
Lead 82	L_α M_α	0.39	Good	[a]	Good	OsL_β	[a] Present as stain in sections

8.5 Problems associated with detecting certain elements

The problems already discussed concerning the effect of specimen preparation, electron irradiation, etc. on the detection of elements in specimens do not apply to all elements in the same way. Each element will have its own peculiar characteristics depending on its state in the specimen being examined. Table 8.1 gives some guidance as to the problems to expect with certain elements. Lighter elements have lower X-ray emission and are more easily absorbed. The stability during preparation or irradiation depends on the type of specimen and on the form of the element in the specimen.

References

Alderson, R. H. (1975), Design of the electron microscope laboratory, in: Practical methods in electron microscopy, A. M. Glauert, ed. (North-Holland, Amsterdam).

Chandler, J. A. (1973), Recent developments in analytical electron microscopy, J. Microscopy *98*, 359.

Appendix

List of suppliers

The following list includes suppliers of materials and instruments necessary for the procedures mentioned in the foregoing chapters. More detailed lists of chemicals for electron microscope specimen preparation are given in the appendices of other books in this series: Glauert 1974 (biological) and Goodhew 1972 (metallurgical and mineralogical). Details of suppliers of materials for cryo-ultramicrotomy are given by Reid (1974). Alderson (1975) gives information for setting up an electron microscope laboratory. This appendix lists mainly those special materials mentioned in the text. For all normal and routine laboratory equipment and chemicals, readers are encouraged to obtain catalogues from the general suppliers listed below.

A monthly list of suppliers of laboratory equipment and materials is published in Laboratory Practice, United Trade Press Ltd., 42–43 Garrard Street, London W1V 7LP, England.

Equipment and materials are also described regularly in The Medical Technologist and Scientist, 30A York Street, Twickenham, Middlesex, England, and International Laboratory, P.O. Box 56522, Rotterdam 3020, The Netherlands. A comprehensive guide to company's names and addresses for electrical, electronic and laboratory products is given in the Product Data Book, Technical Indexes Ltd., Ascot, Berkshire SL5 7EU, England. An index to scientific equipment is also given in *Science* No. 194, edition 4267A, 1976.

The author will be pleased to hear of names and addresses of other suppliers for inclusion in later editions.

A. General suppliers of materials and equipment for electron microscopy and microanalysis

(a) **Agar Aids for Electron Microscopy**
(Agar Aids)
(UK agents for Ladd)
127a Rye Street
Bishop's Stortford
Hertfordshire
England

(b) **Ernest F. Fullam Inc.** (EFFA)
P.O. Box 444
Schenectady
New York 12301
U.S.A.

Graticules Ltd.
(UK agents for EFFA)
Sovereign Way
Tonbridge
Kent, TN9 1RN
England

Touzart & Matignon
(French agents for EFFA)
3 Rue Amyot
75 Paris 5
France

(c) **EMscope Laboratories Ltd.** (EMscope)
374 Wandsworth Road,
London SW8 4TE
England

(d) **Ladd Research Industries Inc.** (Ladd)
(US agents for Agar Aids)
P.O. Box 901
Burlington
Vermont 05401
U.S.A.

(e) **Polaron Equipment Ltd.** (Polaron)
(U.K. agents for Polysciences)
60/62 Greenhill Crescent
Holywell Estate
Watford
Hertfordshire
England

Ted Pella Inc.
(U.S. agents for Polaron)
P.O. Box 510
Tustin
California 92680
U.S.A.

(f) **Polysciences Inc.** (Polysciences)
(U.S. agents for Polaron)
Paul Valley Industrial Park
Warrington
Pennsylvania 18976
U.S.A.

(g) **Taab Laboratories** (Taab)
52 Kidmore End Road
Emmer Green
Reading
Berkshire
England

Extech International Corp. (Extech)
(U.S. agents for Taab)
177 State Street
Boston
Massachusetts 02109
U.S.A.

(h) **E. M. Zairyo-Shya**
308 Yotsuya Sango Building
2 Saemoncho Shinijuku-ku
Tokyo
Japan

(i) **Walter McCrone Associates Inc.**
(McCrone)
2820 South Michigan Avenue
Chicago
Illinois 0616
U.S.A.

(j) **Denton Vacuum Inc.** (Denton)
Cherry Hill Industrial Centre
Cherry Hill
New Jersey 080342
U.S.A.

(k) **SPI Supplies** (SPI)
Division of Structure Probe Inc.
P.O. Box 342
West Chester
Pennsylvania 19380
U.S.A.

(l) **Tousimis Research Corp.** (Tousimis)
6000 Executive Rockville
Maryland 20852
U.S.A.

(m) **Balzers Union AG** (Balzers)
Postfach 75
FL-9496 Balzers
Fürstentum
Liechtenstein

also

Balzers High Vacuum Ltd.
North Bridge Road
Berkhamstead
Hertfordshire
England

also

Balzers High Vacuum Corp.
P.O. Box 10816
Sant Ana
California 92711
U.S.A.

(n) **CW French Inc.** (French)
58 Bittersweet Lane
Weston
Massachusetts 02193
U.S.A.

(o) **Electron Microscope Sciences** (EMS)
P.O. Box 251
Fort Washington
Pennsylvania 19034
U.S.A.

(p) **Vaughn Electron Microscope Supplies Inc.**
2176 Dunn Road
Memphis
Tennessee 38114
U.S.A.

(q) **Electron Microscope Aids**
6 Lime Trees
Malford
Chippenham
Wiltshire
England

B. Materials, chemicals and laboratory equipment for specimen preparation

1. *General laboratory chemicals*

Aldrich Chemical Co. Ltd. (formerly Ralph N. Emanuel Ltd.)
264 Water Road
Wembley
Middlesex HAO 1BR
England

Auriema Ltd.
422 Bath Road
Slough SL1 6BB
Buckinghamshire
England

British Drug Houses Ltd. (BDH)
Poole
Dorset, BH12 4NN
England

Boehringer Corp. (London) Ltd.
Bell Lane
Lewes
East Sussex BN7 1LG
England

Fisher Scientific Co. (Fisher)
633 Greenwich Street
New York
N.Y. 10014
U.S.A.
(U.K. agent: Kodak Ltd., see B21)

Cambrian Chemicals Ltd.
Beddington Farm Road
Croydon CRO 4XB
Surrey
England

E. Merck
61 Darmstadt
West Germany

Hughes and Hughes Ltd.
Elms Industrial Estate
Church Road
Harold Wood
Romford
Essex RM3 OHR
England

Polysciences (*see* A)

Baird and Tatlock
P.O. Box 1
Romford RM1 HA
Essex
England

Eastman Kodak Company
Rochester
New York, 14650
U.S.A.

Hopkins and Williams Ltd.
Chadwell Health
Essex RM1 1HA
England

Sigma (London) Chemical Co. Ltd.
Norbiton Station Yard
Kingston-upon-Thames
Surrey KT2 7BH
England

Koch-Light Laboratories Ltd.
Colnbrook
Buckinghamshire, SL3 OBZ
England

SAS Scientific Chemicals Ltd.
Victoria House
Vernon Place
London WC1B 4DR
England

Fisons Scientific Apparatus
Bishop Meadow Road
Loughborough LE11 ORG
Leicestershire
England

May & Baker Ltd.
Dagenham
Essex RM10 7XS
England

2. *Potassium pyroantimonate*

BDH (*see* B1)
Fisher (*see* B1)

3. *Atomiser for powder spraying*

General suppliers

4. *Pasteur pipettes*

General suppliers

5. *Micropipettes*

Balzers (*see* A)
McCrone (*see* A)

Shandon Southern Instruments Ltd.
Frimley Road
Camberley
Surrey GU16 5ET
England

AHS United Kingdom
Station Road
Didcot
Berkshire
England

Camlab Ltd.
Nuffield Road
Cambridge CB4 1TH
England

Jencons (Scientific) Ltd.
Mark Road
Hemel Hemstead
Hertfordshire
England

6. *Microscope slides*

General suppliers

7. *Diamond knives*

Agar Aids (*see* A)
Balzers (*see* A)
Polaron (*see* A)
Ladd (*see* A)

Friedrich Dehmer
8202 Bad Aibling
Kolpingstrasse 8
West Germany

DuPont de Nemours (France) S.A.
9 Rue de Vienne
Paris 8
France

EI DuPont de Nemours & Co. (Inc.)
Instruments Division
Wilmington
Delaware 19898
U.S.A.

DuPont Co. (UK) Ltd.
Wilbury House
Wilbury Way
Hitchin
Hertfordshire
England

Akashi Seisakusho Ltd.
C.P.O. Box 1405
Tokyo
Japan

Ernst Leitz
Wetzlar
West Germany

GE-FE-RI
146/G Via Marittima I
Frosinone
Italy

Rawyler & Wolf
Pieterlen
Bern
Switzerland

IVIC
Apartitado 1827
Caracas
Venezuela

Rondikin Corp.
2003 Kalia Road
PH-7, Honolulu 96815
U.S.A.

Reichert-Jung U.K.
820 Yeovil Road
Slough
Buckinghamshire
England

8. *Electron microscope grids*

Normal (mesh; slit; finder)

General suppliers

Gilder Grids
23 Macfarlane Road
London W12 7LA
England

Mason & Morton Ltd.
Fir Tree House
Headstone Drive
Wealdstone
Harrow
Middlesex HA3 5QS
England

Smethurst Highlight Ltd.
Bolton
Lancashire
England

Nylon

Agar Aids (*see* A)
EFFA (*see* A)
Graticules (*see* A)
EMscope (*see* A)

Aluminium

Agar Aids (*see* A)
EFFA (*see* A)
Graticules (*see* A)
Polaron (*see* A)
EMscope (*see* A)

Beryllium

Agar Aids (*see* A)
EFFA (*see* A)
EMscope (*see* A)

9. *Carbon rods for evaporation*

General suppliers

Morganite Carbon Co. Ltd.
Battersea Works
London SW11
England

Johnson Matthey Chemicals Ltd.
74 Hatton Garden
London EC1.
England

10. *Metals for evaporation*

General suppliers

Johnson Matthey Metals Ltd.
81 Hatton Garden
London EC1
England

Cambridge Instruments Ltd.
Moat Lane
Melbourne
Royston
Hertfordshire SG8 6EJ
England

11. *Formvar*

General suppliers

Shawinigan Ltd.
118 Southwark Street
London SE1
England

G. T. Gurr
Searle Scientific Services
Coronation Road
Cressex Industrial Estate
High Wycombe
Buckinghamshire
England

12. *Collodion*

General suppliers

13. *Double-sided SEM mounting tape*

General suppliers

14. *Carbon and silver conducting paste*

Agar Aids (*see* A)
EFFA (*see* A)
Polaron (*see* A)
SPI (*see* A)
Tousimis (*see* A)
Balzers (*see* A)
Baird & Tatlock (*see* B1)
EMscope (*see* A)

15. *Diamond cutting wheels*

Cambridge Instruments Ltd. (*see* B10)
Melbourne
Royston
Hertfordshire
England

Shandon (*see* B5)

16. *Beryllium inserts for specimen holders, and beryllium machinery*

Royal Ordnance Factory
Caerphilly Road
Cardiff
Wales

17. *Ion-beam etching instruments*

Edwards Instruments Ltd.
Manor Royal
Crawley
Sussex
England

Alba
Asnieres
Seine
France

Commonwealth Sci. Corp.
500 Pendleton Street
Alexandria, Virginia 22314
U.S.A.

GV Planes Ltd.
Windmill Road
Sunbury on Thames
Middlesex
England

Materials Research Corporation
RT 303
Orangeburg
New York 10962
U.S.A.

18. *Vacuum coating and sputtering units*

AEI (*see* E1)
Agar Aids (*see* A)
Balzers (*see* A)
Denton (*see* A)

Edwards High Vacuum
Manor Royal
Crawley
Sussex
England

EFFA (*see* A)
EMscope (*see* A)

General Engineering
Vacuum Products Division
Station Works
Bury Road
Radcliffe
Manchester M26 9UR
England

JEOL (*see* E1)
Ladd (*see* A)

Leybold Heraeus Ltd.
Blackwell Lane
London SE10 OAU
England

Materials Sciences (NW) Ltd.
55 Cocker Street
Blackpool
Lancashire
England

Nanotech (Thin Films) Ltd.
172 Church Road
Manchester
England

NGN Ltd.
Kirk Road
Church
Accrington
Lancashire
England

Technics
5510 Vine Street
Alexandria
Virginia 22310
U.S.A.

Vacuum Generators
East Grinstead
Sussex
England

Varian AG
Steinhauserstrasse 6300
Zug
Switzerland

Varian Vacuum Division
611 Hansen Way
Palo Alto
California 94303
U.S.A.

19. *Critical point-drying apparatus*

Agar Aids (*see* A)
Balzers (*see* A)
Denton (*see* A)
Ladd (*see* A)
Ted Pella (*see* A)
Polaron (*see* A)
Polysciences (*see* A)
Tousimis (*see* A)

The Bomar Co.
P.O. Box 225
Tacoma
Washington 98401
U.S.A.

Parr Instrument Co.
211 Fifty-third Street
Moline
Illinois 61265
U.S.A.

20. *Buffer solutions*

General suppliers

G. T. Gurr (Gurr)
Searle Scientific Services
Coronation Road
Cressex Industrial Estate
High Wycombe
Buckinghamshire
England

Fisons Scientific Apparatus (Fisons) (*see* B1)

Eastman Kodak
Distillation Products
New York, N.Y.
U.S.A.

21. *Fixatives for electron microscopy* (see Glauert 1974)

General suppliers

Johnson Matthey Chemical Ltd. (*see* B9)

Fisher Scientific (Fisher) (*see* B1)

Gurr (*see* B20)
Fisons (*see* B20)

Shell Chemicals Ltd.
Shell House
Downstream Buildings
London SE1
England

Shell Chemical Corp.
Industrial Chemical Division
415 Madison Avenue
New York, N.Y.
U.S.A.

Kodak Ltd.
(U.K. agents for Fisher)
Chemical Division
Kirkby
Liverpool
England

Union Carbide Corp.
270 Park Avenue
New York, N.Y. 10017
U.S.A.

22. *Embedding materials* (see Glauert 1974)

General suppliers

R.P. Cargille Laboratories Inc.
Cedar Grove
New Jersey 07009
U.S.A.

Ciba-Geigy (Ciba U.K.)
Duxford
Cambridge
England

Ciba Co., Inc. (Ciba U.S.)
Plastics Division, Kimberton,
Pennsylvania
U.S.A.

Ciba Products Corp.
Fair Lawn
New Jersey
U.S.A.

Dow Chemical Co.
Midland
Michigan 48640
U.S.A.

Dow Chemical Europe S.A.
Alfred Escher Strasse 39
8207, Zurich
Switzerland

Fluka AG
Chemische Fabrik
Buchs SG
Switzerland

Okan Shoji Co.
311 Kobikikan 7
6-chome
Ginza-higashi
Chuo-ku
Tokyo
Japan

Marblette Corp.
Long Island City
New York
U.S.A.

Rohm & Haas Co.
Washington Square
Philadelphia
Pennsylvania
U.S.A.

Maumee Chemical Co.
2 Oak St.
Toledo
Ohio
U.S.A.

Pennsalt Chemical Corp.
Three Penn. Center
Philadelphia
Pennsylvania
U.S.A.

Fabriek van Chemische Producten
Vondelingenplaat NV
P.O. Box 7120
Rotterdam
The Netherlands

Thiokol Chemical Corp.
Trenton
New Jersey
U.S.A.

Pittsburg Plate Glass Co.
Plastic Sales
Paint and Brush Division
Pittsburgh
Pennsylvania
U.S.A.

B.I.P. Chemicals Ltd.
Oldbury
Birmingham
England

Société des Usines Chimiques
Rhône-Poulenc
France

Riken Goseijushi Co.
3, 6-chome
Ginza 6-3
Chuo-ku
Tokyo
Japan

Martin Jaeger
Vésenaz
Geneva
Switzerland

Eastman Organic Chemicals
Rochester 3
New York 14650
U.S.A.

Wallace & Tiernan Inc.
Lucido Division
Buffalo
New York 14240
U.S.A.

23. *Dehydration agents and intermediate solvents* (see Glauert 1974)

General suppliers

Fisons (*see* B20)
Gurr (*see* B20)
Union Carbide (*see* B21)

24. *Osmometer*

Knaur, K.G. Dr. Ing.
Herbert Knaur & Co GmbH
1 Berlin 37
Holstweg 18
West Germany

A. D. Whitehead
The Ancient House
Ardleigh
Colchester
Essex
England

25. *Freeze-drying apparatus*

Edwards (*see* B17)
EMscope (*see* A)

Meditronics Inc.
P.O. Box 11209
Dallas
Texas
U.S.A.

Chemlab Instruments Ltd.
602 High Road
Ilford
Essex
England

26. *Dichlorohexyl-18-crown-6 for making resin standards*

Aldrich (*see* B1)
Polysciences (*see* A)

27. *Carbon and beryllium planchets for SEM specimen mounting*

Agar Aids (*see* A)
EFFA (*see* A)
SPI (*see* A)
Ladd (*see* A)
EMscope (*see* A)

Madison Engineering Co.
P.O. Box 927
Columbia
Maryland 21044
U.S.A.

McCrone (*see* A)

28. *General SEM mounting stubs*

General suppliers

Madison (*see* B27)

29. *Adhesives for mounting SEM specimens*

General suppliers

30. *Replicating material*

General suppliers

BX Plastics Ltd.
Manningtree
Essex
England

31. *General histology materials*

Astell Laboratory Service Co.
172 Brownhill Rd.
Catford
London SE6 2DL
England

Cambridge Instruments Ltd. (*see* B10)

Difco Laboratories (UK)
P.O. Box 14B
Central Avenue
West Molesey
Surrey
England

Degenhardt & Co. Ltd.
31–36 Foley Street
London W1P 8AP
England

Raymond A. Lamb
6 Sunbeam Road
London NW10 6JL
England

32. *Adhesive for X-ray window mountings*

Bostik Ltd.
Leicester
England

General suppliers

33. *Micromanipulators*

Carl Zeiss Jena Ltd.
CZ Sci. Inst. Ltd.
93/97 New Cavendish Street
London W12 AR
England

Electrotech Associates
Prince of Wales Industrial Estate
Abercarn
Monmouth
England

Scientific Techniques Ltd.
Reliant Works
Brockham
Surrey
England

Research Instruments Ltd.
Ace Works
Cumberland Avenue
London NW10
England

Ealing Beck Ltd
Greycaine Road
Watford WD2 4PW
Hertfordshire
England

Unimatic Engineers Ltd.
Granville Road Works
122 Granville Road
London NW2
England

Micro Instruments (Oxford) Ltd.
7 Little Clarendon Street
Oxford OX1 2HP
England

E. Leitz (Instruments) Ltd.
48 Park Street
Luton
Bedfordshire
England

34. *Plasmod low temperature ashing equipment*

Agar Aids (*see* A)
Nanotech (*see* B18)

35. *Ultrasonic baths*

General suppliers

Ultrasonic Ltd.
Otley Road
Shipley
Yorkshire
England

Kerry Ultrasonics Ltd.
Wilbury Way
Hitchin
Hertfordshire
England

C. Microtomes and Ultramicrotomes

1. *Ultramicrotomes* (see Reid 1974)

Cambridge Huxley Mk 2

Cambridge Instruments Ltd. (*see* B10)

LKB Ultratome 1, III and IV, and LKB Huxley

LKB Instruments Ltd.
232 Addington Road
South Croydon, CR2 8YD
Surrey
England

Reichert OMU2, OMU3

C. Reichert Ltd.
Hernalser Hauptstrasse 219
A-1171
Wien
Austria

Reichert-Jung U.K. (*see* B 7)

Sorvall Porter Blum MT1, MT2, MT2B

Du Pont Co.
Instrument Products Division
Sorvall Operations
Newton
Connecticut 06470
U.S.A.

V. A. Howe & Co. Ltd.
88 Peterborough Road
London SW6
England

Leitz Ultramicrotome

E. Leitz Inc.
468 Park Avenue South
New York, N.Y. 10016
U.S.A.

E. Leitz (Instruments) Ltd.
30 Mortimer Street
London W1
England

or

48 Park Street
Luton
Bedfordshire LU1 3HP
England

2. *Microtomes*

Astell Laboratory Service Co. (*see* B31)
Cambridge Instruments Ltd. (*see* B10)
Degenhardt & Co. Ltd. (*see* B31)
Raymond A. Lamb (*see* B31)
E. Leitz (*see* C1)
C. Reichert Ltd. (*see* C1)
Du Pont Ltd. (*see* B7)

D. Cryotechniques

1. *Cryo-ultramicrotomy* (*see* Reid 1974)

Cryokit for LKB Ultratome I, III and IV
LKB (*see* C1)

FC150 Freezing attachment for Reichert OMU3
C. Reichert (*see* C1)

Frozen thin sectioner attachment for FTS/LTC-2 for Sorvall Porter-Blum MT2B and MT2
Sorvall (*see* C1)

SLEE cryo-ultramicrotome
SLEE (*see* D2)

2. *Cryostats*

SLEE (South London Electrical Equipment Ltd.)
179 Lanier Road
Hither Green Lane
London SE13 6QD
England

E. Leitz (Instruments) Ltd. (*see* C1)

Bright Instrument Co. Ltd.,
Clifton Road
Huntingdon PE18 7EU
England

3. *Quenching media*

Isopentane
BDH (*see* B1)

Methyl cellulose
BDH (*see* B1)

Propane 7
BDH (*see* B1

Freon 12, *Freon* 22*:* Balzers AG (*see* A)

J. T. Baker Chemical Co.
Phillipsburg
New Jersey
U.S.A.

Arcton

ICI Ltd.
Mond Division
Rocksavage Works
Runcorn
Cheshire
England

4. *Chemicals for cryo-ultramicrotomy*

Gelatin

EFFA (*see* A)
Polysciences (*see* A)
Taab (*see* A)
Fisher (*see* B1)

Dimethyl sulphoxide (DMSO)

General suppliers

Fisher (*see* B1)

Bovine serum albumin (BSA)
EFFA (*see* A)
Gurr (*see* B11)
Polysciences (*see* A)
Taab (*see* A)

Miles Laboratories Inc.
Kanake III
Illinois 60901
U.S.A.

Ethylene glycol

Fisons (*see* B1)
Fisher (*see* B1)

E. Electron-optical instruments and accessories

1. *Electron-optical and analytical instruments*

Table A.1 indicates the current range of instruments available, showing which models have facilities for X-ray analysis (marked with an asterisk). Where instruments are adopted from other configurations they are indicated by the symbol ★. Instruments having field emission guns are indicated by (f). Addresses of the manufacturers and their agents are given below.

(a) **AEI Scientific Apparatus Ltd.** (AEI)
Barton Dock Road
Urmston
Manchester M31 2LD
England

AEI Scientific Apparatus Inc.
Kratos
4075 Ruffin Road,
San Diego
California 92123
U.S.A.

(b) **Advance Metals Research Corporation** (AMR)
160 Middlesex Turnpike
Bedford
Massachusetts
U.S.A.
(U.K. agents: **E. Leitz** (C1))

(c) **Applied Research Laboratories** (ARL)
9545 Wentworth Street
Sunland
California 91040

or
Wingate Rd.
Luton
Bedfordshire, England

(d) **Cambridge Instruments Ltd.**
(CSI)
Melbourne
Royston
Hertfordshire SH8 6EJ
England

Cambridge Instrument Co. Inc.
40 Robert Pitt Drive
Monsey
New York 10952
U.S.A.

(e) **Cameca**
103 Boulevard St. Denis
Courbevoie (Seine)
France

U.K. agents: **AEI**

U.S. agents:
Cameca Instruments Inc.
101 Executive Boulevard
Elmsford
New York 10523
U.S.A.

(f) **Carl Zeiss**
7082 Oberkochen
West Germany

Carl Zeiss
31–36 Foley Street
London W1P 8AP
England

Carl Zeiss Inc.
444 Fifth Avenue
New York, N.Y. 10018
U.S.A.

(g) **Coates & Welter Instrument Corp.**
(Coates & Welter)
777 North Pastoria Avenue
Sunnyvale
California 94086
U.S.A.

U.K. agents:
Reichert-Jung U.K. Ltd.
(*see* B7)

TABLE A.1 (for key see E1 on p. 531)

Manufacturer	TEM	SEM	STEM	EPMA
AEI	EM801 EMMA-4* CORINTH 275* CORINTH 500* CORA*	CESA ★	STEM 1176*(f)	
AMR		AMR 1200* AMR 1000* AMR 1000A* AMR 1000FE(f)* AMR 1400(f)*	AMR 1000*★ AMR 1000A*★ AMR 1000FE(f)	
ARL		SEMQ*★	SEMQ*★	SEMQ*
CSI		Stereoscan 600* Stereoscan 150* Stereoscan 180*	Stereoscan 600*★ Stereoscan 150*★ Stereoscan 180*★	Microscan 9*
Cameca		MBS 100*		MBX*
Carl Zeiss	EM9 S-2 EM 10A* EM 10B*	Novascan 30*		
Coates & Welter		50 A*(f) 102 A*(f) 104 A*(f) 106 A*(f) 30 B*(f) 102 B*(f) 104 B*(f) 106 B*(f) 2000*(f)★	50 A*★(f) 102 A*★(f) 104 A*★(f) 106 A*★(f) 30 B*★(f) 102 B*★(f) 104 B*★(f) 106 B*★(f)	
ETEC		Autoscan*	Autoscan*★(f)	

Hitachi	HS9-1 HS9-2 H500 A* H500 ST* H700*	H500 A*★ H500 ST*★ H700*★ S500* S550* S700(f)* HFS-2(f)*	H500 A*★ H500 ST*★ H700*★ S500*★ S700(f)*★ HFS-2(f)*★	
ISI		Mini SEM MSM 7* Super I* Super II* Super III*	MSM 7*★ Super I*★ Super II*★ Super III*★	
JEOL	JEM 100C* JEM 120C* JEM 100S* JEM 200B*	JEM 100C*★ JEM 120C*★ JEM 100S*★ JEM 200B*★ JSM 35* JFSM 30*(f) P15*	JEM 100C*★ JEM 120C*★ JEM 100S*★ JEM 200B*★ JXA 50A*★ JSM 35*★ JFSM 30*(f)★	JXA 50A
MAC	MAAK-1	700* 700A*	700A*★	MAC-5 450* 400S*
Philips	EM 400* EM 301* EM 201	EM 400*★ EM 301*★ PSEM 500*	EM 400*★ EM 301*★ PSEM 500*★	
Siemens	Elmiskop CT150* Elmiskop CT100F*(f) Elmiskop 101* Elmiskop 102* Elmiskop 1*	Elmiskop 101★* Elmiskop 102★* Autoscan*(f)	ST 100F*(f) Elmiskop 101★* Elmiskop 102★*	Autoprobe*
VG		HB 50*(f)	HB 5*(f) HB 50*(f)★	

(h) **ETEC Corporation** (ETEC)
3392 Investment Boulevard
Hayward
California 94545

U.S.A.
European agents:
Siemens AG

(i) **Hitachi Ltd.**
Tokyo
Japan

U.S. agents:
Perkin–Elmer Corporation
Instruments Division
Main Avenue
Norwalk
Connecticut 06856
U.S.A.

U.K. agents:
Perkin–Elmer Ltd.
Post Office Lane
Beaconsfield
Buckinghamshire HP9 1QA
England

(j) **International Scientific Instruments Inc.**
(ISI)
Suite 5
San Antonio Road
Palo Alto
California 94303
U.S.A.

ISI
6050 Offenbach am Main
Dreichring 48
West Germany

ISI
Waterwitch House
Exeter Road
Newmarket
Suffolk
England

(k) **Japan Electron Optical Laboratories**
(JEOL)
148 Nakagami Akishima
Tokyo 196
Japan

JEOLCO (UK) Ltd.
JEOLCO House
Grove Park
Edgeware Road
Colindale
London NW9
England

JEOL
477 Riverside Avenue
Bedford
Massachusetts 02155
U.S.A.

(l) **Materials Analysis Company** (MAC)
1060 East Meadow Circle
Palo Alto
California 94303
U.S.A.

(m) **N.V. Philips** (Philips)
Gloeilampenfabrieken
Eindhoven
Netherlands

U.K. agents:
Pye Unicam Ltd.
York St.
Cambridge CB1 2PX
England

U.S. agents:
Philips Electronic Instruments
750 South Fulton Avenue
Mount Vernon
N.Y. 10550
U.S.A.

(n) **Siemens Aktiengesellschaft**
Wernewerk für Messtechnik
7500 Karlsruhe 21
Rheinbrückestrasse 50
West Germany

Siemens (U.K.) Ltd.
Great West House
Great West Road
Brentford
Middlesex TW8 9DG
England

Siemens of America Inc.
350 Fifth Avenue
New York
N.Y. 10001
U.S.A.

(o) **Vacuum Generators Ltd. (VG)**
The Birches Industrial Estate
Imberhorne Lane
East Grinstead
Sussex
England

Vacuum Generators GmbH
1 Berlin 52
Schillingstrasse 29
West Germany

U.S.A. agents:
58 Buckingham Drive
Stamford
Connecticut 06902
U.S.A.

2. *Instrumental accessories*

(a) *Pointed filaments*

EBTEC
CW French Division
5 Shawsheen Avenue
Bedford
Massachusetts 01730
U.S.A.

(b) *Makrofol and other plastic films for thin X-ray windows*

Siemens (*see* E1)

Yarsley Research Laboratories Ltd.
Clayton Road
Chessington,
Surrey
England

(c) *Biological ultrathin standards*

Agar Aids (*see* A)

(d) *Field emission guns*

Coates & Welter (*see* E1)

(e) *Lanthanum hexaboride cathodes*

Agar Aids (*see* A)

Kimball Physics Inc.
Wilton
New Hampshire 03086
U.S.A.

(f) *Microanalysis multielement standards*

Charles M. Taylor Co.
Stanford
California
U.S.A.

Tousimis (*see* A)
Agar Aids (*see* A)

(g) *Apertures*

General suppliers

(h) *Normal EM filaments*

General suppliers

(i) *Closed circuit water circulators*

Agar Aids (*see* A)

Eyles Refrigeration Co.
Dunkirk Mills
Woodchester
Stroud
Gloucestershire
England

EMscope (*see* A)

F. X-ray equipment

1. *X-ray detectors and energy analysers*

(a) **Link Systems Ltd.**
Halifax Road
High Wycombe
Bucks HP12 35E
England

(b) **Canberra Industries Inc.**
45 Gracey Avenue
Meriden
Connecticut 06450
U.S.A.

(c) **Edax International Inc.**
P.O. Box 135
Prairie View
Illinois 60069
U.S.A.

Edax Europe BV
Nassaulaan 5
Den Haag
Netherlands

(d) **Geoscience Nuclear**
Division of Geoscience Inst. Corp.
2335A Whitney Avenue
Hamden
Connecticut 06518
U.S.A.

(e) JEOL (*see* E1)

(f) **Kevex Corporation**
Analytical Instruments Division
898 Mahler Road
Burlingame
California 94010
U.S.A.

(g) **Nuclear Equipment Corporation**
963A Industrial Road
San Carlos
California 94070
U.S.A.

(h) **Nuclear Enterprises Ltd.**
Sighthill
Edinburgh EH11 4EY
England

(i) **Nuclear Semiconductors**
163 Constitution Drive
P.O. Box 2367
Mento Park
California 94025
U.S.A.

(j) **Ortec**
Materials Analysis Division
100 Midland Road
Oak Ridge
Tennessee 37830
U.S.A.

U.K. agents:
Ortec Ltd.
Dalroad Industrial Estate
Dallow Road
Luton
Bedfordshire LU1 1SU
England

European agents:
Ortec GmbH
8 München 13
Frankfurter Ring 81
West Germany

(k) **Princeton Gamma-Tech**
Box 641
New Jersey 08540
U.S.A.

(l) **Siemens AG**
ZVW 104
D-800 München 1
Postfach 103
West Germany

(m) **Tracor Northern**
2551 West Beltline Highway
Middleton
Wisconsin 53562
U.S.A.

Tracor Europe BV
Schiphol Airport Amsterdam
Building 106
The Netherlands

2. *X-ray spectrometers and crystals*

Microspec Corp.
265-G Sobrante Way
Sunnyvale
California 94086
U.S.A.

Nuclear & Silica Products Ltd.
44–46 The Green
Wooburn Green
High Wycombe
Buckinghamshire
England

Quartz & Silice SA
8 Rue d'Anjou
75 Paris 8
France

Q.B.I. International
2034 Golden Gate Avenue
San Francisco
California 94115
U.S.A.

Tousimis (*see* A)

3. *Gas-flow proportional counters*

MAC (*see* E1)
Tousimis (*see* A)

G. Analytical laboratories

EBTEC Corp.
5 Shawsheen Avenue
Bedford
Massachusetts 01730
U.S.A.

EMV Associates Inc.
Microanalysis Laboratory
15825 Shady Grove Road
Rockville
Maryland 20850
U.S.A.

Le Mont Scientific
1359 East College Avenue
State College
Pennsylvania 16801
U.S.A.

Micron Inc.
P.O. Box 3136
Wilmington
Delaware 19807
U.S.A.

Mid America Microanalysis Laboratories Inc.
4379 South Howell Avenue
Milwaukee
Wisconsin 53207
U.S.A.

McCrone (*see* A)

Photometrics Inc.
442 Marret Road
Lexington
Massachusetts 02173
U.S.A.

Scanatlanta Research Corp.
1645 Tully Circle, NE
Suite 101, Atlanta
Georgia 30329
U.S.A.

Scanservice Ltd.
82 High Street
Gt. Missenden
Buckinghamshire
England

St. Louis Scanning Electron Microscopy Laboratory
P.O. Box 672
Bridgeton
Missouri
U.S.A.

Structure Probe Inc.
535 East Gay Street
West Chester
Pennsylvania 19380
U.S.A.

or:

230 Forrest Street
Metuchen
New Jersey 08840
U.S.A.

Tousimis (*see* A)

International Laboratory Services (ILS)
99 New Cavendish Street
London WIM 7FQ
England

TAAB (agents) (*see* A)
EMscope (agents) (*see* A)
Agar Aids (agents) (*see* A)

References to Appendix

Alderson, R. H. (1975), Design of the electron microscope laboratory, In: Practical methods in electron microscopy, A. M. Glauert, ed. (North-Holland, Amsterdam).

Glauert, A. M. (1974), Fixation, dehydration and embedding of biological specimens, In: Practical methods in electron microscopy, A. M. Glauert, ed. (North-Holland, Amsterdam).

Goodhew, P. J. (1972), Specimen preparation in materials science, In: Practical methods in electron microscopy, A. M. Glauert, ed. (North-Holland, Amsterdam).

Reid, N. (1974), Ultramicrotomy, In: Practical methods in electron microscopy, A. M. Glauert, ed. (North-Holland, Amsterdam).

Index for list of suppliers

Subject index